Native North-Eastern Ornamental Fish Breeding Techniques

THE AUTHORS

Dr. Mangesh M. Bhosale had his Ph.D with Senior Research Fellowship at the Aquatic Rainbow Technology Park (ARTP), Department of Aquaculture, Fisheries College & Research Institute, Tamil Nadu Dr. J. Jayalalithaa Fisheries University, Ponneri from 2016-2019. He has carried out applied research in advanced aquariculture production systems, fish reproduction and live food production systems. He has published many research papers in national and international journals. He received his B.F.Sc candidature from College of Fishery Science, MAFSU, Udgir and Master degree (M.F.Sc., Fish Biotechnology) from College of Fisheries, Dr. BSKKV, Ratnagiri.

Prof. S. Felix has 38 years of teaching and research experience in aquaculture. He is the Past President of World Aquaculture Society-Asian Pacific Chapter and he has also served as Director of the World Aquaculture Society(APC). Prof. S. Felix has served as Second Vice Chancellor of Tamil Nadu Dr. J. Jayalalithaa Fisheries University and also served as Dean of the Fisheries College in Tamilnadu, India. He has developed 40 Constituent units for the University including half a dozen new Fisheries Institutes to impart fisheries education and trainings. He developed and introduced the Raceway culture technology for the first time in India. Under his effective leadership the Aquatic Rainbow Technology Park and an Aquatic Rainbow Mall for ornamental fish culture were developed at Chennai, as first of its kind in the country. He has 10 patents published on various aquaculture technologies to his credit. He has published 15 Text Books, 50 Manuals and more than 150 Research papers and reports.. Introduced 30 new courses comprising UG, PG ,Ph.D and Paraprofessional programs. Implemented 30 R&D and Technology projects to develop a dozen aquaculture technologies for the benefit of farmers.

Native North-Eastern Ornamental Fish Breeding Techniques

Mangesh M. Bhosale
S. Felix

2021
Scholars World
A Division of
Astral International Pvt. Ltd.
New Delhi – 110 002

Disclaimer:

Every possible effort has been made to ensure that the information contained in this book is accurate at the time of going to press, and the publisher and author cannot accept responsibility for any errors or omissions, however caused. No responsibility for loss or damage occasioned to any person acting, or refraining from action, as a result of the material in this publication can be accepted by the editor, the publisher or the author. The Publisher is not associated with any product or vendor mentioned in the book. The contents of this work are intended to further general scientific research, understanding and discussion only. Readers should consult with a specialist where appropriate.

Every effort has been made to trace the owners of copyright material used in this book, if any. The author and the publisher will be grateful for any omission brought to their notice for acknowledgment in the future editions of the book.

Cataloging in Publication Data--DK

Courtesy: D.K. Agencies (P) Ltd. <docinfo@dkagencies.com>

Bhosale, Mangesh M., author.

Native north-eastern ornamental fish breeding techniques / Mangesh M. Bhosale, S. Felix.

pages cm

Includes bibliographical references.

ISBN 9789354616341

1. Ornamental fishes--Breeding--India, Northeastern. 2. Fish culture--India, Northeastern. I. Felix, S., author. II. Title.

LCC SF457.B46 2021 | DDC 639.34 23

Published by : **Scholars World**
A Division of
Astral International Pvt. Ltd.
– ISO 9001:2015 Certified Company –
4736/23, Ansari Road, Darya Ganj
New Delhi-110 002
Ph. 011-4354 9197, 2327 8134
E-mail: info@astralint.com
Website: www.astralint.com

Preface

Fisheries sector now accord a place of importance to aquaculture. In many Asian countries, aquaculture has been a way of life for farmers from centuries, its status in the context of aquatic resource management and employment generation in rural areas remained until recently a matter of debate. The scenario has changed radically and the spectacular success of certain types of aquaculture enterprises such as aquariculture(Ornamental fish culture). Agencies all over the world considered it as a major area, and several scientific and technical institutions are presently involved in research on various aspects. All these have contributed to the recognition of some of the basic needs and problems of this new and emerging sector. The present information highlighted the importance of the development and improvement of technology through research and dissemination of information for the rapid and orderly development of the sector.

Many present-day practices are based on traditional techniques with only limited involvement of recent technologies. This major handicap is now being increasingly understood, and the sector has been recognized as a major science. Live food production, which is the backbone of aquaculture and especially for aquariculture is yet to be developed for its applications. To access the existing experience and information is extremely difficult. It is believed that commercial application of the present technologies will result in much greater production, only in the presence sufficient numbers of trained and experienced personnel. The world conferences, symposiums, workshops, and many later international consultations, have repeatedly emphasized the crucial importance of training to develop innovative strategies to conserve the nature. The key personnel are technicians at the middle management level, with a broad background of the more important disciplines involved and practical experience of the relevant technologies. They may not be experts in all the disciplines, but their background will help in executing aquariculture development. While establishing Aquatic Rainbow Technology Park (ARTP) project at Chennai, we came to recognize the need for a comprehensive book covering the various aspects involved in aquariculture for educational as well as technology development programs.

This book is an attempt to fill this need, and to present the future scene of the aquaculture sector. As the main objective of the book is to develop breeding and larval rearing culture systems for North East endemic ornamental fish species in advanced aquaculture systems that presently contribute to the production, or are likely to do so in the near future. Though all the major aspects are covered, this book is essentially intended for those who are eagerly willing to conserve the nature and to obtain a basic understanding of the principles and practices of the sector. As the emphasis is on practical applications, exhaustive reviews of research are not attempted, although major findings of importance are included.

Authors

Contents

Preface *v*
List of Tables *xv*
List of Figures *xvii*
List of Plates *xix*

1. Introduction **1**

2. Review of Literature **5**

2.1 Systematic position of fish species investigated 5
2.2 Brief description of the species 5
2.2.1 Moustached danio (*Danio dangila*) 5
2.2.2 Sophore barb (*Puntius sophore*) 5
2.3 Induced breeding of ornamental fishes at globe and India 6
2.4 Status of induced breeding of North-East ornamental fishes 7
2.5 Role of synthetic hormones in induced breeding 7
2.6 Systematic position of fresh water Rotifer 10
2.7 Role of live food in aquaricuture 10
2.8 Technologies for rotifer production 11
2.9 Role of micro-algae on rotifer production 12
2.10 Technologies on algal (*Chlorella vulgaris*) production 12
2.11 Role of probiotics in larval rearing 14
2.12 Effect of probiotic incorporated feed on larval health 14
2.13 Raceway technology for larval rearing 15
2.14 Effect of stocking densities on larval rearing 15

3. Materials and Methods **17**

3.1 Induced breeding of *Danio dangila* and *Puntius sophore* 17
3.1.1 Sample collection and experimental site for animals 17

3.1.2 Physico-Chemical parameters of water 22

3.1.3 Selection of brooders 23

3.1.4 Setting of spawning tank 24

3.1.5 Procurement of hormone and preparation 24

3.1.6 Induced breeding 24

3.1.7 Breeding performance 27

3.1.8 Embryonic and post-embryonic stages 28

3.2 Mass production of fresh water Rotifers (***Brachionus calyciflorus***) in outdoor culture system 29

3.2.1 Procurement and preparation of Rotifer (*Brachionus calyciflorus*) stock 29

3.2.2 Procurement and preparation of *Chlorella vulgaris, Spirulina major* and *Scenedismus obliquus* cultures 29

3.2.3 Experimental setup 30

3.2.4 Sampling of Rotifers 32

3.2.5 Feeding intensity 32

3.3 Mass production of *Chlorella vulgaris* 34

3.3.1 Procurement of *C. vulgaris* and media for culture 34

3.3.2 Design of internal photobioreactor 34

3.3.3 Control of Ciliates 34

3.3.4 Measurement of *Chlorella vulgaris* biomass and specific growth rate 35

3.3.5 Determination of protein content in *C. vulgaris* 35

3.3.6 Determination of lipid content in *C. vulgaris* 36

3.3.7 Determination of pigment content 36

3.4 Mass production of freshwater Rotifers (*Brachionus calyciflorus*) in indoor culture system 37

3.4.1 Sourcing of Rotifers (*Brachionus calyciflorus*) 37

3.4.2 Experimental design 37

3.4.3 Sampling of Rotifers 37

3.4.4 Estimation of algal ingestion rate by rotifer 39

3.4.5 Mass culture and harvesting protocol 39

3.4.6 Maintenance of Algal culture 40

3.4.7 RAS (Recirculatory Aquaculture System) facility for continuous culture of rotifer 40

3.5 Formulation of probiotic incorporated encapsulated feed 45

3.5.1 Isolation, identification and characterization of *Lactobacillus sp.* from selected fish intestine as a probiotic 45

3.5.2 Feed preparation 49

3.5.3 Encapsulation of probiotic bacteria in feed 50

3.5.4 LAB occurrence test 52

3.5.5 Effect of probiotic incorporated encapsulated feed on larval growth 52

3.5.6 Histological analysis of selected fish species 53

3.6 Larval rearing of selected ornamental fishes 53

3.6.1 Effect of rearing system and stocking density on larval rearing 53

3.7 Statistical analysis 57

4. Results 59

4.1 Induced breeding of Moustached danio, *Danio dangila* (Hamilton, 1822) 59

4.1.1 Breeding behavior 59

4.1.2 Latency period 59

4.1.3 Number of eggs spawned 59

4.1.4 Fertilization rate 60

4.1.5 Hatching rate 60

4.2 Induced breeding of Sophore barb, *Puntius sophore* (Hamilton, 1822) 65

4.2.1 Breeding behavior 66

4.2.2 Latency period 66

4.2.3 Number of eggs spawned 66

4.2.4 Fertilization rate 66

4.2.5 Hatching rate 66

4.3 Mass production of Rotifer, *Brachionus calyciflorus* in outdoor culture system 72

4.3.1 Effect of different food types on the growth of Rotifer, *B. calyciflorus* 72

4.3.2 Effect of feeding interval on rotifer, *B. calyciflorus* production 73

4.3.3 Water quality parameters 74

4.4 Mass production of single cell algae, *Chlorella vulgaris* 75

4.4.1 Effect of light wavelengths on ***Chlorella vulgaris*** production 75

4.4.2 Effect of different light wavelengths at 5k lux on cell size, protein and pigment enhancement in *C. vulgaris* 76

4.4.3 Control of Ciliates 78

4.5 Mass production of Rotifers, *B. calyciflorus* in indoor RAS based system 79

4.5.1 Effect of different food type on rotifer production in indoor culture system 79

4.5.2 Estimation of algal, *Chlorella vulgaris*, consumption rate by individual rotifer, *Brachionus calyciflorus* 80

4.5.3 Effect of daily harvest rate on Rotifer production: 81

4.5.4 Effect of Recirculatory Aquaculture system (RAS) on Rotifer production 82

4.6 Formulation of probiotic incorporated encapsulated feed 85

4.6.1 Isolation, identification and characterization of *Lactobacillus* ***sp.*** from fish intestine 85

4.6.2 Preparation and characterization of probiotic incorporated encapsulated feed 89

4.6.3 Effect of probiotic incorporated feed on larval growth 90

4.6.4 Histological analysis of selected fish species 91

4.7 Larval rearing of *Danio dangila* and *Puntius sophore* 94

4.7.1 Effect of rearing system and stocking density on *Danio dangila* 94

4.7.2 Effect of rearing system and stocking density on *Puntius sophore* 95

5. Discussion 103

5.1 Induced breeding of North-East region based ornamental fishes, Moustached Danio, *Danio dangila* and Sophore barb, *Puntius sophore* in captivity 103

5.1.1 Induced breeding of Moustached Danio, *Danio dangila* 103

5.1.2 Induced breeding of Sophore barb, ***Puntius sophore*** 105

5.2.1 Mass production of freshwater Rotifer, *Brachionus calyciflorus* in outdoor culture system 107

5.2.2 Biomass production and pigment enhancement of single cell algae, ***Chlorella vulgaris*** in internal photobioreactor 108

5.2.3 Mass production of freshwater Rotifer, *Brachionus calyciflorus* in indoor culture system 112

5.3 Formulation of Probiotic incorporated encapsulated feed 115

5.3.1 Isolation, identification and characterization of *Lactobacillus sp.* from fish gut 115

5.3.2 Preparation and characterization of probiotic incorporated encapsulated feed 117

5.3.3 Effect of probiotic incorporated encapsulated feed on larval growth of *D. dangila* and *P. sophore* 118

5.4 Larval rearing of *Danio dangila* and *Puntius sophore* 119

5.4.1 Effect of rearing system and stocking density on ***Danio dangila*** 119

5.4.1 Effect of rearing system and stocking density on ***Puntius sophore*** 120

6. Summary and Conclusion **123**

7. References **127**

List of Tables

Table 1:	Reports on captive breeding of fishes using pituitary gland extract, synthetic hormones and environmental manipulation	8
Table 2:	Morphological characters of broodstock in ***Danio dangila***	23
Table 3:	Morphological characters of broodstock in ***Puntius sophore***	23
Table 4:	Treatments based on hormonal dose applied to induce fishes	27
Table 5:	Treatments based on different feed type used in culture of freshwater rotifer in outdoor culture system	32
Table 6:	Treatments based on light wavelengths for biomass production and pigment enhancement of *Chlorella vulgaris* in internal photobioreactor	34
Table 7:	Effect of common purified salt on control of ciliates in algae, *C. vulgaris*, culture tanks	35
Table 8:	Treatments based on different feed type used in the culture of freshwater rotifer in indoor culture system	37
Table 9:	Treatments based on algal cell concentration on ingestion and filtration rate of freshwater rotifer, ***Brachionus calyciflorus***	39
Table 10:	Effects of different harvesting rates on rotifer production in RAS based system	40
Table 11:	Ingredients used for feed preparation and proximate composition	50
Table 12:	Effect of probiotic incorporated encapsulated feed on selected ornamental fish larvae	52
Table 13:	Effect of rearing system and stocking density on larval rearing of selected ornamental fish species	57
Table 14:	Effect of different hormones and dosages on reproductive parameters of ***Danio dangila*** during induced breeding	60
Table 15:	Description on embryonic development of *Danio dangila* recorded during captive breeding experiment (Plate 49)	63
Table 16:	Effect of different hormones and doses on reproductive parameters of Puntius sophore during induced breeding	67
Table 17:	Description on embryonic development of Puntius sophore recorded during captive breeding experiment (Plate 50)	69

Table 18:	Effect of different food types on Rotifer, ***B. calyciflorus*** production in outdoor culture system	73
Table 19:	Effect of feeding interval (2× and 3× day^{-1}) on Rotifer production in outdoor culture system	74
Table 20:	Effect of light wavelengths on cell concentration and specific growth rate of *Chlorella vulgaris*	76
Table 22:	Effect of common purified salt on control of Ciliates in algal, *C. vulgaris*, culture tanks	78
Table 23:	Effect of different food type on Rotifer (*B. calyciflorus*) production in indoor culture system (Rotifers mL^{-1})	79
Table 24:	Estimation of algal cell, *Chlorella vulgaris*, consumption rate by Rotifers in Jar (500 mL)	81
Table 25:	Effect of daily harvest on Rotifer production in Recirculatory Aquaculture System (RAS)	82
Table 26:	Water quality parameter and rotifer numbers (range) observed during study period	83
Table 27:	Colony morphology of isolated bacteria (11 colonies) from *D. dangila* intestine (Plate 36 A and B).	85
Table 28:	Colony morphology of isolated bacteria (09 colony) from *P. sophore* (Plate 37 A and B)	85
Table 29:	Biochemical tests of isolated colonies from *D. dangila* intestine	87
Table 30:	Biochemical tests of isolated colonies from *P. sophore* intestine	87
Table 31:	Antibiotic susceptibility test of isolates from *D. dangila* and *P. sophore* (Plate 38 A and B)	88
Table 32:	pH and bile salt tolerance for *D. dangila* and *P. sophore* isolates	89
Table 33:	LAB viability test after encapsulation (24 hours).	90
Table 34a:	Effect of probiotic incorporated feed on growth indices of *D. dangila* larvae	91
Table 34b:	Effect of probiotic incorporated feed on growth indices of *P. sophore* larvae	91
Table 35:	Histological analysis of intestine of *D. dangila* and *P. sophore* larvae	93
Table 36:	Growth performance of ***D. dangila*** larvae reared in two different systems	95
Table 37:	Growth performance of *P. sophore* larvae reared in two different systems.	99

List of Figures

Figure 1:	Effect of different hormones and dosages on mean latency period (hours) in *D. dangila*	61
Figure 2:	Effect of different hormones and dosages on mean number of eggs spawned (eggs female^{-1}) in *D. dangila*	62
Figure 3:	Effect of different hormones and dosages on mean fertilization rate (%) in *D. dangila*	62
Figure 4:	Effect of different hormones and dosages on mean hatching rate (%) in *D. dangila*	62
Figure 5:	Effect of different hormones and dosages on mean latency period (hours) in *P. sophore*	68
Figure 6:	Effect of different hormones and dosages on mean number of eggs spawned (eggs female^{-1}) in *P. sophore*	68
Figure 7:	Effect of different hormones and dosages on mean fertilization rate (%) in *P. sophore*	69
Figure 8:	Effect of different hormones and dosages on mean hatching rate (%) in *P. sophore*	69
Figure 9:	Effect of different food type on Rotifer, *B. calyciflorus* production in outdoor culture system	73
Figure 10:	Effect of feeding interval (2× and 3× day-1) on Rotifer production in outdoor culture system	74
Figure 11:	Water quality parameters in 2× feeding intensity tanks	75
Figure 12:	Water quality parameters in 3× feeding interval tanks	75
Figure 13:	Effect of different light wavelengths at 5k lux on *C. vulgaris* production in internal photobioreactor	76
Figure 14:	Effect of different light wavelengths at 5k lux on protein and pigment enhancement in *C. vulgaris* produced using internal photobioreactor	77
Figure 15:	Effect of different food type on Rotifer production in indoor culture system (Rotifers mL^{-1})	80

Figure 16: Algal cell, *Chlorella vulgaris*, consumption rate by Rotifers 81

Figure 17: Effect of daily harvesting rate on Rotifer production in Recirculatory Aquaculture System (RAS) 82

Figure 18: Monthly Rotifer production in RAS system. 83

Figure 19: Dissolved oxygen, pH and ammonia during culture period 84

Figure 20: Alkalinity and Hardness during culture period 84

Figure 21: Temperature variation during culture period of rotifer in RAS system 84

Figure 22: Effect of stocking density on survival rate (%) of Danio dangila in nursery raceways 96

Figure 23: Effect of stocking density on final weight gain (g) of *Danio dangila* in nursery raceways 96

Figure 24: Effect of stocking density on average daily gain (g) of *Danio dangila* in nursery raceways 97

Figure 25: Effect of stocking density on specific growth rate (%) of Danio dangila in nursery raceways 97

Figure 26: Effect of stocking density on survival rate (%) of *Puntius sophore* in nursery raceways 99

Figure 27: Effect of stocking density on final weight gain (g) of *Puntius sophore* in nursery raceways 100

Figure 28: Effect of stocking density on average daily gain (g) of *Puntius sophore* in nursery raceways 100

Figure 29: Effect of stocking density on specific growth rate (%) of *Puntius sophore* in nursery raceways 100

List of Plates

Plate 1:	Loktak Lake, Moirang, Manipur	17
Plate 2:	Brahmaputra stretch, Rajabari, Assam	18
Plate 3:	Fish sample collection	18
Plate 4:	Ornamental fish hatchery, ICAR sub-center for NEH region, Manipur	19
Plate 5:	Fish packing at Manipur	19
Plate 6:	Moustached danio, *Danio dangila*	20
Plate 7:	Sophore barb, *Puntius sophore*	20
Plate 8:	Multi-species ornamental fish hatchery, TNJFU, Madhavaram	21
Plate 9:	Acclimatization of collected ornamental fishes	22
Plate 10:	Rearing of collected wild stock in RAS - FRP tanks	22
Plate 11:	*Danio dangila* broodstock A. Male and B. Female	24
Plate 12:	*Puntius sophore* broodstock A. Male and B. Female	24
Plate 13:	Experimental setup for induced breeding of selected ornamental fishes	25
Plate 14:	Intra-Peritonial injection to *Danio dangila* and *Puntius sophore*	26
Plate 15:	Eggs released at the bottom of tanks after inducement	27
Plate 16:	Counting of fertilized eggs	28
Plate 17:	Observation of *Chlorella vulgaris* under microscope (40X)	30
Plate 18:	*Chlorella vulgaris* colonies on agarose	30
Plate 19:	Stock culture of *Chlorella vulgaris*	31
Plate 20:	Mass culture of *Chlorella vulgaris*	31
Plate 21:	Live feed production unit (outdoor)	32
Plate 22:	Experimental setup for outdoor rotifer production	33

Plate 23:	Rotifer sample collection and observation	33
Plate 24:	Water quality analysis	33
Plate 25:	Live feed production unit (FRP based)	34
Plate 26:	Layout and set up of internal photobioreactor with LEDs in culture tanks of C. *vulgaris*	36
Plate 27:	RAS based live food production unit	38
Plate 28:	Experimental setup for live food and algae culture	38
Plate 29:	Rotifer sample collection and observation	40
Plate 30:	Algal, *Chlorella vulgaris,* culture tanks with under water light facility	41
Plate 31:	Layout of Live food production unit with RAS facility	42
Plate 32:	Cartridge and U.V. filtration setup (in-series)	43
Plate 33:	Sand filter, Air blower and water pump attached at life food unit	43
Plate 34:	Rotifer, *Brachionus calyciflorus,* culture tanks with under wateraeration facility	44
Plate 35:	Collection of intestine from *Danio dangila* and *Puntius sophore*	45
Plate 36:	A - Isolated Lactobacillus sp. colonies from *D. dangila*	47
	B - Screened colonies on MRS from *D. dangila*	47
Plate 37:	A - Isolated *Lactobacillus sp.* colonies from *P. sophore*	47
	B - Screened colonies on MRS from *P. sophore*	47
Plate 38:	Antibiotic susceptibility test of isolates: **A.** *D. dangila* and **B.** *P.* sophore	49
Plate 39:	Stock preparation of *Lactobacillus spp.*	49
Plate 40:	Mass production of *Lactobacillus spp.*	51
Plate 41:	Probiotic incorporated encapsulated feed with different concentrations	51
Plate 42:	LAB occurrence test	53
Plate 43:	Spawns of one day old	54
Plate 44:	Spawn of three day old	54
Plate 45:	RAS based nursery raceways	55
Plate 46:	Filtration facility (sand and U.V. filters) at nursery raceway	55

Plate 47:	Rearing of *Danio dangila* in nursery raceways	56
Plate 48:	Rearing of *Puntius sophore* in nursery raceways	56
Plate 49:	Embryonic developmental stages recorded during captive breeding of *D. dangila*	65
Plate 50:	Embryonic developmental stages recorded during captive breeding of *P. sophore*	72
Plate 51:	Major Ciliates observed in algal, *C. vulgaris,* culture tanks	78
Plate 52:	Histological appearance of *Danio dangila*	92
Plate 53:	Histological appearance of *Puntius sophore*	92
Plate 54:	Histological examination of intestinal villi in *D. dangila.*	93
Plate 55:	Histological examination of intestinal villi in *P. sophore.*	93
Plate 56A:	Abnormal villi	
Plate 56B:	Normal villi	94
Plate 57:	Larval development of *Danio dangila*	98
Plate 58:	Larval development of *Puntius sophore*	101

CHAPTER – 1

Introduction

Globally there is a steady increase in the ornamental fish trade with an annual growth of 14% due to the enhanced popularity (MPEDA, 2016 and Dey, 2016). Keeping colorful fishes as pets in aquaria or garden pools is an age old hobby which was originated in China by keeping of gold fish in a glass bowl (Kelly, 1987). During 17th century, gold fish was introduced to several parts of the world and became popular in most of the countries and hence ornamental fish trade has started as an occupation initially later on as a profession (Bhattacharjya and Choudhury, 2004). The first public display aquarium was opened at Regent Park in England in 1853 (Swain et al., 2003). Presently there are over 500 display aquaria functioning worldwide. In India, the hobby of keeping ornamental fishes as pets is of recent origin with the opening of the Taraporewala Aquarium at Mumbai in 1951 (Kurup et al., 2004). In Northeastern region, aquarium keeping as a hobby got a boost with the setting up of first aquarium shop in Guwahati during 1977 (Bhattacharjya and Choudhury, 2004). Recently, Aquatic Rainbow Technology Park (ARTP) has been established by Tamil Nadu Dr. J. Jayalalithaa Fisheries University in Chennai, Tamil Nadu for up scaling ornamental fish production and promoting incredible indigenous ornamental fishes at international level.

Around 500 million fishes are traded annually in the world by 145 countries of which tropical species contribute 85% share (Gouvaia et al., 2003). The trade in ornamental fishes is an important source of income for many countries (Saxby et al., 2010 and Ploeg, 2013). Northeast India (21° 57′N to 29° 30′N Longitude and 89° 46′E to 97°30′E Latitude) is one of the 34th mega biodiversity hotspots listed by Conservation International, the Himalayas and Indo-Burma (Roach, 2005) and is endowed with a vast expanse of fresh water habitats mainly flood plains (1.59 Lakh ha); rivers (20,875km); reservoirs (0.33 lakh ha); tanks and ponds (3.71 lakh ha) including the mighty Brahmaputra and the adjoining wetlands (Nath, 2014). Being one of the highest precipitated zone in the world, the hilly states of Arunachal, Nagaland, Manipur, Mizoram, Tripura and Meghalaya are having innumerable streams, lakes and a variety of aquatic habitats (Bijoya et al., 2014). Sandilyan (2016) reported that India is endowed with rich fish diversity having around 600 ornamental fish species of which North-East India is blessed with 296 potential ornamental fish species contributing up to 11.6% of total freshwater species (genera: 114; families: 38 and orders: 10) (Vishwanath, 2010 and Dhar and Ghosh, 2015).

Freshwater fish family Cyprininae includes barbs, with about 3000 living and extinct species but only 1270 remain extant, which are characteristic in having large scales, bright colors, schooling behavior and easy to maintain made them popular in the aquarium trade (Clyde *et.al.*, 1997). Aquatic biodiversity is being depleted at an alarming rate due to habitat destruction, pollution, overexploitation due to high export demand, tourism and introduction of invasive exotic species along with alien pathogens and parasites (Sreevalli et al., 2014; Convention on Biological Diversity, 2014). In general, the loss is higher in the freshwater ecosystem, which supports 40% of the globally recorded fish species. Out of 250 ornamental fish species identified under vulnerable category, more than 10 species are critically endangered while 28 are under endangered category (Dey et al., 2016). Till date, captive breeding for only 13 species has been succeeded from North-East Indian region (Dey et al., 2016).

In aquaculture, larval culture is a main bottleneck in promoting higher production and there is no efficient diet for initial feeding of fish larvae (Felix, 2012). Thus, the hatcheries need efficient live feed at right time (Felix, 2013 and Rojo-Cebreros et al., 2017). Rotifers, *Brachionus calyciflorus* (Phylum Rotifera) are small (approximately 200 micrometers, μm) metazoan and have been widely used as essential live food source in raising aquaculture due to its unique characteristics like easily digestible, swims slowly, has appropriate size, survives in high density (Odo et al., 2015 and Gilbert, 2017) and apposite biochemical composition (Lubzens et al., 2001). Moreover, it has the potential for enrichment with fatty acids, vitamins and various therapeutants to reduce huge loss during earlier developmental stages and for production of healthy fish respectively (Maragelman et al., 1985). In addition, they have the habit of staying suspended in the water column, high reproduction rate and high density cultures with temperature tolerance of 15 to 31°C at an optimal pH 6-8 (Sananurak et al., 2009).

With the remarkable development of aquaculture sector, demand for Rotifers has been increasing considerably (Ashraf et al., 2010 and Rojo-Cebreros et al., 2017). But, Rotifers experience extremes of variations with respect to abiotic and biotic factors in natural waters (Nogrady *et al.* 1993). Accordingly, various investigations has been undertaken for getting higher production in smaller areas with recirculatory facilities, identification of appropriate food and for maintaining the ecosystem at its constant during live food production in various parts of the world. Regrettably, the sustainable rotifer production technology is still at a reverie juncture. Thus, making hatcheries as a successful venture can be attributed partly, if not totally, by successful mass cultivation of rotifers in sustainable way. In general, not much attention has been given on its culture in freshwater and in our local environment in specific. Intense efforts are needed in this minuscule to make our fish hatcheries as a successful venture by supplying abundant quantity of rotifers.

Fish larvae are the smallest self-supporting vertebrates and in order to increase their chances of survival, they need to complete morpho-functional systems to escape from predation and obtain sufficient quantity of food (Sandilayan, 2016). The transition from endogenous to exogenous feeding is one of the most critical stages in freshwater fish culture and the availability of suitable food is essential for high density success on commercial scale of aquariculture (Sasmita, 2016). The requirement of nutrients varies throughout the life cycle of an individual fish. At early stages, the requirement of nutrients is comparatively high. The difficulties in larval rearing are primarily related to nutrition and secondly, the small mouth size (Dhert et al., 1998). Live foods such as rotifers, Cladocerans and brine shrimp nauplii have been used for a long time as first foods. The failure to do this usually leads to high larval mortalities (Odo et al., 2015).

Diet particles must be chosen with consideration to the small mouth size of larvae. The selection of very small feed particles leads to nutrient leaching problems caused by very high ratio of surface area to volume (Kim et al., 1996). In order to design a good larval diet, even before considering the nutrient requirements of the larvae, another basic factor must be considered like the feed must be ingested by the larvae and should be digested. Probiotics play an important role in promoting digestion and maintaining animal health (Borch et al., 2015). The encapsulation techniques for protection of bacterial cells have resulted in enhancing digestibility in the animals (Chavarri et al., 2012). In addition, encapsulation can promote controlled release and optimize delivery to the site of action (Kato et al., 2016).

In aquaculture, term raceway is a highly generic and implies little more than a water impoundment with water flow-through and Recirculatory Aquaculture system (RAS). The Romans and perhaps even earlier cultures - used complex water management systems to effectively create and manage both aquaculture raceways and static pond systems. The evolution of the linear aquaculture raceway started with at least as far back as the Romans. After reviewing known examples chronologically, it is easy to see the evolutionary path of the stacked raceway and its related variations. Many of the stacked raceway refinements and variations have come from many aquaculture species and even other non-aquaculture elements of modern industrial agricultural food production and even including sewage processing technology.

The rearing of the larvae to the fry stage is most critical in the cycle of fish seed production in hatcheries and its rearing under controlled hatchery condition requires the development of specific culture techniques (Olaniyi and Omitogun, 2013). Using conventional systems such as smaller tanks, tubs, etc. for ornamental fish farming is becoming outdated (Felix, 2014). On the other hand, research developments in larvae culture and early rearing technology have allowed 90% of currently marketed fresh water ornamental fish to be cultured (Tlusty, 2002). In order to produce more and to make more profits, advanced culture systems must be adopted. Raceway farming would enable the sector to grow at a different level in our country. Raceways have various advantages over other culture techniques like work efficiently with high stocking density, can be managed at zero water exchange, high survival, easy to manage, suitable for mass culturing of various ornamental fishes, higher growth with clean water and economically viable (Felix, 2008 and 2013). Hence, in present study, nursery and growout rearing was done in raceways for mass production of selected ornamental fish species.

Hence, the present study was initiated, to conserve as well as to meet the demand of national and international market, to develop mass production technology of North-East region based ornamental fishes, Moustached Danio

(*Danio dangila,* Hamilton, 1822, least concerned, IUCN, 2018), and Sophore barb (*Puntius sophore,* Hamilton, 1822, least concerned, IUCN, 2018) locally known as *Giant danio* and Chebli, respectively, as no reports are available on natural or induced breeding of these species. After mass production, larval development is another critical stage due to unavailability of efficient diet. To overcome this bottle neck, the study on mass production technology of Rotifers, *B. calyciflorus* in both outdoor and indoor RAS based culture system was undertaken. The nutrient requirement of fish larvae varies according to development and supply of live feed throughout the development may fails sometime. Hence, a kind of diet is required to fulfill the nutritional requirements. Therefore, the succeeding objective was commenced to formulate probiotic incorporated encapsulated diet followed by systematizing larval rearing technique in advance culture systems like raceways.

Objectives:

i) Standardizing induced breeding technique for mass production of North-East origin based ornamental fishes *Danio dangila* and *Puntius sophore*

ii) Mass scale production technology for fresh water rotifer (*Brachionus calyciflorus*)

iii) Formulation of probiotic incorporated encapsulated feed for selected ornamental fish species

iv) To standardize larval rearing technique for selected ornamental fish species using advanced aquaculture system like raceways

CHAPTER – 2

Review of Literature

2.1 Systematic position of fish species investigated

Kingdom : Animalia

Subkingdom : Bilateria

Phylum : Chordata

Subphylum : Vertebrata

Infraphylum : Gnathostomata

Superclass : Actinopterygii (ray-finned fishes)

Class : Teleostei

Superorder : Ostariophysi

Order : Cypriniformes (cyprins, minnows)

Superfamily : Cyprinoidea

Family : Cyprinidae

Genus : ***Danio***

Species: ***Danio dangila*** (Hamilton, 1822)

Genus : ***Puntius***

Species: ***Puntius sophore*** (Hamilton, 1822)

Conservative status: Least concerned (IUCN, 2018)

2.2 Brief description of the species

2.2.1 Moustached danio (*Danio dangila*)

Danio dangila (Hamilton, 1822) locally called Moustached danio (Angami, 2012) and is an endemic fish inhabits mountain streams belonging to Ganga-Brahmaputra drainage (Talwar and Jhingran, 1991). The species have brilliant coloration of body with brownish to olive on back and silvery yellowish in belly and flanks. Three to four silvery bluish lines on flanks and the bands of the body break up anteriorly forming a mottled pattern, which make the fish more attractive and unique among the *Danio* species. Upper angle of gill opening occasionally shows a dusky spot. The species becomes mature at 5-6 cm but can attain a length of about 12 cm (Dahanukar, 2010). These characteristics put *Danio dangila* as classified ornamental fish (Vishwanath, 2010).

2.2.2 Sophore barb (*Puntius sophore*)

Puntius sophore (Hamilton, 1822) locally called Spotfin swamp barb (Mahapatra,

2010) is an indigenous fish inhabits plains and submontane regions of North East India especially from mountain streams belonging to Ganga-Brahmaputra drainage (Talwar and Jhingran, 1991). It is very plentiful shoaling fish, appreciably smaller, beautiful with silver body and back is grey-green to brownish, flanks with bluish luster, a deep black round blotch at the base of the caudal fin and on central part of dorsal fin, a long red line on anterior part of the body make fish as very beautiful animal in aquarium keeping. The species becomes mature at 4-6 cm but can attain a length of about 13 cm (Dahanukar, 2010). Recent field surveys showed that population in the wild is continuously declining in extent of occurrence due to exploitation for ornamental fish trade (Gerstner et al., 2006 and Dey, 2016).

2.3 Induced breeding of ornamental fishes at globe and India

A hobbyist or entrepreneur would naturally like to know more about beautiful ornamental fishes like its habitat, feeding behavior, compatibility, courtship altogether known as the breeding behavior and its techniques. About 24,600 fish species have been reported from a wide range of habitats worldwide. Among these, more than 600 species become popular as ornamental fishes (Sajeevan, 2013). Freshwater fish family (Cyprinidae) includes 2,400 barbs in about 220 genera which are characteristic in having large scales, bright colours, schooling behaviour and are easy to maintain which made them popular in the aquarium trade (Clyde et al., 1997; Biswas et al., 2015 and Van, 2017).

Notable contributions have been made by Hamilton (1822), Sands (1986), Kelly (1987), Mills (1990), Nelson (1994), Riehl and Baensch (1996), Vinci (1998), Vishwanath (2010) and Sandilayan (2016) towards cataloging and recording the worldwide distribution of tropical freshwater ornamental fishes. On the culture and breeding of exotic ornamental fishes, important contributions were made by Plona (1962), Anderson (1965), Kaufman (1965), Fernado and Phang (1985), Andrew (1986) and Polanski (1991).

However, the aquatic habitats especially in the developing countries have witnessed rapid changes in their environment that has caused greater habitat degradation during past few decades (Geist and Hawkins, 2016). Like in other parts of the world, the fish habitats of our part of the globe also degraded as a consequence of natural and ever increasing human interferences (Umesh et al., 2012 and Stevens et al., 2017).

Induced spawning in fish, employing hypophysation as a tool was first developed by the Brazilians, (Von Ihering, 1937). In India, Khan (1938) was the first to attempt induced breeding in Indian major carps (IMC) by using mammalian pituitary. The first successful induced breeding of carp was reported in India, way back by Choudhuri and Alikunhi in 1957. The most commonly used method for stimulation of maturation is the administration of fish pituitary extract (Kozlowski 1994). However, this simple method is not always successful. In many cases, much better results in controlled breeding were obtained when pure or synthetic forms of hormones are used for reproductive stimulation (Takashima et al., 1984; Crim et al., 1987; Drori et al., 1994 and Motilan et al., 2014).

Significant contributions are made by several workers from last century (*Von Ihering*, 1937; Choudhuri and Alikunhi, 1957; Sinha, 1972; Dixit and Agarwal, 1974; David and Rahman, 1975; Chaco and Kuriyan, 1984; Barua and Mollah, 1987; Akteruzzaman et al., 1991; Mahapatra, 1999; Abidi and Thakur, 1997; Sarkar and Ponniah, 2000; Chaudhury and Biswas, 2003; Dey and Sarmah 2000, 2003; Sarmah and Dey, 2004; Sarmah, 2002, 2003; Mitra, 2004; Swain, 2008; Swain and Singh, 2008, Mangesh et al., 2017 and Kumar et al., 2018) on the development of standard breeding technologies for ornamental fishes.

2.4 Status of induced breeding of North-East ornamental fishes

North-East India is mega diversity hot spot, but very few reports are available on successful breeding of ornamental fishes. In recent years, focus has been shifted towards developing the culture and breeding technologies for ornamental fishes especially from degrading hotspots (Sandilayan, 2016). Several workers (Dey, 1995; Das, 1989; Nath and Dey, 2000; Sen, 2000; Dey et al., 2002; Bhattacharjya et al., 2004; Mahapatra, 2004; Vishwanath, 2010; Goswami et al., 2012; Bijoya et al., 2014; Dhar and Ghosh, 2015 and Sandilayan, 2016) have also made valuable contribution on the fish diversity of North-East India, which has motivated to work on the ornamental fish species of the region.

Artificial induction spawning behavior in *Puntius ganionotus* (Bleeker) was done by Liley and Tan (1895) while Vincent and Thomas (2008) studied nuptical colcuration and courtship behavior during induced breeding of *Puntius chola*. Prakash et al., (2014) has developed captive breeding and larval rearing technology of *Danio aequipinnatus*. Angami (2012) has discussed repoductive biology of *Danio dangila* and *P. chola* from North-East region of India. In similar manner, Motilan *et al.* (2014) has successfully bred *Puntius manipurensis* using WOVA-FH at the doses of 0.4 mL kg^{-1} resulting highest egg output (6218.75±32.75). Mahapatra (2010) has succeeded in induced breeding and rearing of indigenous ornamental fish *P. shalynius* using ovaprim @ 0.4mL kg^{-1}. In similar manner, Udit *et al.* (2014) used ovatide at a dose of 1.0 mL kg^{-1} for induced breeding of *Puntius sarana*. Volschenk et al., (2018) has successfully bred *Synodontis zambezensis* and studied embryonic development adapting a variety of protocols previously used in breeding of *Clarias gariepinus*, *Synodontis petricola* and *Synodontis nigromaculatus*.

2.5 Role of synthetic hormones in induced breeding

Various environmental factors are known to play important role in the release of gonadotropic hormones (Motilan *et. al.*, 2013). However, the optimum environmental condition for natural breeding is not available in the aquarium at all places. Hence, most of the fish lose their natural breeding behavior when they get migrated from their locality to another geographical spot for rearing (Motilan et al., 2014). Fish culturist's thirst of having an induced breeding agent was satisfied by the development of hormonal use either as a gonadotropin (fish or mammalian, pure or crude) or a GnRH analogue (GnRHa & HCG) with or without a dopamine antagonist (Harvey and Carolsfeld, 1993; Bhattacharya *et. al.*, 2002 and Mangesh et al., 2016).

The time of first successful attempt of induced breeding of ten-spotted live bearer, *Cresterdon decemmaculatus* (Jenyns, 1842) was undertaken by injection of pituitary extract, till then the search for better methods for induced spawning got momentum (Houssay, 1931). Mammalian gonodotropins and adreno-corticosteroids have been used to spawn a variety of fishes and obtained variable results (Billard et al., 1983; Hossain et al., 2008 and Sasmita, 2016).

Ovarian steroid hormones play important role in oocyte growth and maturation in teleosts. In number of teleosts, Oestradiol-17b (20 P - dihydroxy-pregnen-3- one) and 17a (20 p - dihydroxy-pregnen) are concerned in the regulation of oocyte growth and maturation (Nagahama, 1987). Kubokawa et al., (1999) reported that, in fish, sex steroid hormone levels become high during the pre-spawning. (So et al., 1985 and Dye *et al,* 1987). Similar results were published in Coho salmon, *Oncorhynchus kisutch* and *O. mykiss* (Fitzpatrick et al., 1986); White suckers, *Castostomus commersoni* (Scott et al., 1984) and Gold fish, *Carassius auratus* (Stacey, et al., 1979; Kagawa, et al., 1983 and Kobayashi, et al., 1999). But there is scarce research work on indigenous ornamental fishes (Sandilayan, 2016).

At present synthetic hormones, with a formulation incorporating salmon GnRH with or without dopamine antagonists like domperidone, metachlopramide, pimozide (ovaprim, ovapel, ovatide, WOVA-FH, HCG etc.) (Harvey and Carolsfed, 1993), has become the favorite for fish breeders around the world. Another natural product, *Nalrum muriaticum,* a homeopathy preparation was attempted in the induced breeding of gold fish, black molly and *Puntius conchonius* and obtained better result (Mitra and Raizada, 1986 and Sudha and Gokula, 2015). Several other studies have been conducted on the induced breeding of fishes, which are listed below in Table 1.

Table 1: Reports on captive breeding of fishes using pituitary gland extract, synthetic hormones and environmental manipulation

Inducing Agent	Fish	Reference
Pituitary gland	*Oncorhynchus kisutch*	Hunter *et al.* (1981)
	Clupea harengus pallasi	Kreiberg *et al.* (1987)
	Aristichthys nobilis	Thalathiah *et al.* (1988)
	Puntius gonionotiis	Thalathiah *et al.* (1988)
	Silurus asotus	Choi *et al.* (1992)
	Tor khudree	Nandeesha *et al.* (1993)
	Mystus aor	Gupta *et al.* (1993)
	Labeo dussumieri	Kurup *et al.* (1993)
	Heterobranchus longifilis	Nwadukwe (1993)
	Puntius gonionotus	Bhuiyan *et al.* (2006)
	Puntius sophore	Islam (2012)

Ovaprim	*Puntius conchonicus*	Kumar et al., (1988)
	Puntius gonionotus	Kumar et al., (1988)
	Puntius japanicus	Azad and Shimary (1991)
	Ictalunis punctatus	Goudie *et al.* (1992)
	Sillago ciliate	Battaglene (1996)
	Ompak bimaculatus	Sridhar *et al.* (1998)
	Clarias striatus	Francis, *et al.* (2000)
	Poecilia sphenops	Vishakan (2002)
	Carrasius auratus	Vishakan (2002)
	Mystus gulio	Alam *et al.* (2006)
	Puntius chola	Vincent and John Thomas (2008)
	Puntius shalynius	Mahapatra (2010)
	Ompok pabda	Purkayastha *et al.* (2012)
	Puntius sarana	Udit *et al.* (2014)
Ovatide	Puntius javanicus	Thakur and Reddy (1997)
	Heteropneustes fossilis	Marimuthu *et al.* (2000)
	Channa striatus	Marimuthu, *et al.* (2001)
	Poecilia sphenops	Vishakan (2002)
	Carassius auratus	Vishakan (2002)
	Puntius chola	Vincent and John Thomas (2008)
	Puntius sarana	Udit *et al.*(2014)
	Puntius conchonius	Sudha and Gokula (2015)
Human Chorionic Gonadotropin (HCG)	Heterobranchus longifilis	Legendre (1986)
	Pangasius hypophthalmus	Legendre *et al.* (1999)
	Pangasius bocourti	Cacot *et al.* (2002)
	Mystus gulio	Alam *et al.* (2006)
	Heterobranchus bidorsalis	Olaniyi and Omitogun (2014)
	Mystus gulio	Kumar *et al.* (2018)
WOVA-FH	Anabas testudeneus	Sarkar *et al.* (2005)
	Thai carp	Sukumaravin *et al.* (2006)
	Puntius manipurensis	Motilan *et al.* (2014)
	Puntius chola	Motilan *et al.* (2014)
	Botia rostrata	Dey *et al.* (2015)
	Botia Dario	Dey *et al.* (2015)
	Anabas testudeneus	Singh *et al.* (2015)
	Barilius barila	Dey *et al.* (2016)
	Hemichromis bimaculatus	Mangesh *et al.* (2017)

Environmental manipulation	*Cobitis taenia*	Juchno and Boron (2006)
	Colisa sota	Mitra *et al.* (2006)
	Trichogaster trichopterus	Bindu *et al.* (2014)
	Puntius sarana	Udit *et al.* (2014)
	Danio aequipinnatus	Prakash *et al.* (2014)
	Trichogaster labiosua	Motilan and Nishikanta (2015)
	Etroplus suratensis	Felix *et al.* (2017)
	Trichogaster lalius	Saha *et al.* (2017)

2.6 Systematic position of fresh water Rotifer

Kingdom : Animalia

Subkingdom : Bilateria

Infrakingdom : Protostomia

Superphylum : Platyzoa

Phylum : Rotifera (Cuvier, 1817)

Class : Monogonta

Subclass : Monogononta (Plate, 1889)

Superorder : Pseudotrocha (Kutikova, 1970)

Order : Ploima (Hudson and Gosse, 1886)

Family : Brachionidae (Ehrenberg, 1838)

Genus : ***Brachionus*** (Pallas, 1766)

Species : ***Brachionus calyciflorus*** (Pallas, 1766)

2.7 Role of live food in aquaricuture

It is well known fact that feed is a vibrant component and major constraint in the aquaculture industry especially during larval rearing of many cultivable and ornamental finfish and shellfish species in hatcheries. Live food organisms (phytoplankton and zooplankton) are being successfully used for rearing the larvae as food and became one of the inexpensive alternatives to commercial feeds. Earlier development stages of most of the species depend on zooplankton due high nutrient requirement (De Pauw et al., 1984; Watanabe et al., 1983 and Bryant and Matty, 1980). Among various zooplankters, the most important groups are rotifers, copepods, *Artemia,* Cyclops and Cladocerans (Arimoro, 2006; Kimura and Tomaru, 2013; Felix, 2014 and Odo et al., 2015).

Rotifers are a group of mostly planktonic pseudocoelomates characterized by possessing a wheel of cilia called corona at the anterior end and posterior foot for locomotion and creating feeding currents (Gilbert, 2017). Rotifers actively graze water column for feeding and ingest small particles (~1-20μ) like algae (Chew and Lim, 2006) with life span of about 12-15 days but can reproduce on 2-5 days onwards parthenogenetically (Arimoro, 2006; Odo et al., 2015 and Onal et al., 2015).

There are considerable reports on feeding successes with high growth and survival rates using the rotifer, *Brachionus sp.* as live food for several fish species (Villegas et al., 1990; Reitan et al., 1993; Ottera 1993, Craig et al., 1994; Lim and Wong, 1997; Castell et al., 2003; Rimmer et al., 1994; Cho et al., 2001 and Mercier et al., 2004). Among the documented information of Lim and Wong, (1997), an overall survival rate of larvae fed on the rotifers ranged between 65.1% and 74.5% in freshwater ornamental fish larvae culture. Similarly, Arimoro and Ofojekwu (2003 and 2004) obtained over 80% survival in the toothed carp, *Aphyosemion gardneri* larvae raised on the freshwater rotifer, *B. calyciflorus*. Accordingly, various researchers documented that, rotifers fed to fry enhanced their growth and survival rates (Craig et al., 1994; Rimmer et al., 1994; Cho et al., 2001; Castell et al., 2003; Shiri et al., 2003; Mercier et al., 2004; Arimoro et al., 2006; Benetti et al., 2008; Sananurak et al., 2009 and Loka et al., 2016).

2.8 Technologies for rotifer production

However, the hatcheries are depending on mass availability of rotifer (Nogrady *et al.* 1993 and Felix, 2014) but till date, rotifers are extensively cultured in batch culture systems using yeast-based products which are characterized by harvesting all rotifers once at a predetermined density level usually within couple of days after start up as higher densities are subject to sudden crash due to poor water quality and limited area (Oeie and Olsen 1993; Sananurak et al., 2009 and Onal et al., 2015).

Semi continuous culture method is also known as "thinning culture", in which rotifer density is kept constant by periodic harvesting. First, micro algae culture is introduced into a tank (similar to the batch culture method) followed by addition of rotifers and feed on commencing days. As the rotifer attains the predetermined density level, part of the culture is harvested and replaced with micro-algae. Contrary to the batch culture method, this long- term culture is maintained at a low density for a period of 7-14 days without water quality treatment as the size of the culture tank is quite larger than in batch culture method. Several studies have been undertaken on developing sustainable semi-continuous culture method for rotifers (Fu et al., 1997; Yoshimura et al., 1997; Suantika et al., 2000; Dhert et al., 2001; Hagiwara et al., 2001; Rombaut et al., 2003; Bentley et al., 2008; Onal et al., 2010 and 2015).

Intensive rotifer culture is the production of live rotifers at the lowest possible cost (Sauntika et al., 2003). This system consists of two stages. The stage I is continuous algal culture using filtered water enriched with nutrients for algal growth and multiplication (James et al., 1988). The stage II is rotifer culture system with periodic harvesting at predetermined density level for maintaining the culture at stationary phase throughout the season by continuous feed with enriched micro-algae (Hirata and Yamasaki, 1983; Boraas and Bennet, 1988; Fu et al., 1997; Walz et al., 1997; Suantika et al., 2001 and 2003; Arimoro, 2005 and Bentley et al., 2008).

Fu et al., (1997) developed an automatic continuous culture system and which consists of filtration unit, a culture unit and a harvest unit. In this system filtered water and food are continuously supplied into a rotifer culture tank at a pre-determined rate, and the same amount of culture water is transferred into a harvest tank to obtain rotifers at a significant biomass. The average production was about 2.1 billion rotifers day^{-1} from a 1 m^3 S type continuous culture in which high rotifer densities ranging from 3000 to 6000 nos. mL^{-1} was maintained.

2.9 Role of micro-algae on rotifer production

Rotifers are filter feeders and can be fed with a variety of food types, including algae, yeast, bacteria or inert foods such as microcapsules and detritus. The type of feed used for culturing rotifers can have a significant effect on the cost of operations and on the nutritional value of rotifers. (Hirayama et al., 1989; Carnic et al., 1993; Reitan et al., 1993; Suantika et al., 2000; Castell et al., 2003; Planas et al., 2004; Prol-Garcia et al., 2010; Aragao et al., 2004; Arimoro, 2006 and Loka et al., 2016). Various mechanisms have been suggested to explain the poor survival of rotifers and one of them is the scarcity of favorite algal species in required quantity and competition for food by other organisms (Oeie and Olsen 1993 and Sananurak et al., 2009).

Micro-algae are the principal component of most cultured rotifer diet. Many species of algae are employed according to availability under local conditions. The most commonly used species are *Nannochloropsis oculata, Scenedismus obliquus, Isochrysis galbana, Chlorella vulgaris, Spirogyra sp., Chlamydomonas reinhardtii, Botryococcus braunii, Neochloris oleoabundans, Haematococcus pluvialis* and *Spirulina major* (Hirayama et al., 1989; Brett and Muller-Navarra, 1997; Nandini and Sarma, 2000, 2001; Ikawa, 2004; Arimoro, 2006; Tomohita et al., 2006; Rajendiran and Subramanian, 2007; Terence et al., 2008; Ngoc-phuong et al., 2009; Probir et al., 2011 and Odo et al., 2015).

Many more new culture methods for rotifer production have been developed enabling high density populations in continuous cultures but all of these culture methods rely on the use of concentrated suspensions of algal species (Abu-Rezq et al., 1997; Fu et al., 1997 and Yoshimura et al., 1997). Similarly, James et al., (1987), Reitan et al., (1993), Suantika et al., (2000), Castell et al., (2003), Planas et al., (2004), Arimoro, (2006), Rajendiran and Subramanian, (2007) and Odo et al., (2015) observed that the micro-algal cell density in rotifer culture not only influence the rotifer population but also the presence of essential amino acids. At algal density of 40.5 x 10^6 cells mL^{-1} of *Chlorella sp.,* there is no inhibition in the biosynthesis of these essential fatty acids, however, decline of *n-3* and *n-3* acids was observed, when algal density was applied at 5x10^6 *Chlorella sp.* cells mL^{-1}.

2.10 Technologies on algal (*Chlorella vulgaris*) production

Chlorella vulgaris is single-cell green algae and belongs to the phylum Chlorophyta. It reaches a size of 5 to 10 μm in diameter and has an almost spherical shape (Scragg et al., 2003). Each individual cell is surrounded by a cell wall and contains one nucleus, one chloroplast, several mitochondria, vacuoles, and starch grains. It is a robust and fast growing algae that tolerates a wide range of environmental conditions (Liang et al., 2009). It thrives in freshwater as well as under brackish water conditions, survives temperatures up to 38°C (Kessler, 1972), and may grow phototrophic, heterotrophic, or mixotrophic. It is this versatility and robustness that made *Chlorella vulgaris* as one of the most cultivated microalgae for commercial applications worldwide (Pulz and Gross, 2004 and Liang et al., 2009).

With the above benefits, *Chlorella vulgaris* is an excellent food for early stages of finfishes, shell fishes and molluscans (Rosenberry, 1998 and Muller-Feuga, 2004). It is also the most important dietary source for most of the zooplankton communities in the wild and contributes to the nutrient supply (Hirayama et al., 1989; Carnic et al., 1993; Reitan et al., 1993; Suantika et al., 2000; Castell et al., 2003; Planas et al., 2004; Prol-Garcia et al., 2010; Aragao et al., 2004; Loka et al., 2016; Arimoro, 2006 and Loka et al., 2016).

There are several factors affecting the growth of *Chlorella vulgaris* are classified into two categories: 1. Nutritional factors (chemical) and 2. Environmental factors (physical). Nutritional factors include the composition and amount of the chemical in culture medium like carbon, nitrogen, phosphorus, silicon, metals such as iron, copper, zinc, vitamins, etc. The type and concentration of the carbon source and nitrogen source are very effective, so they are considered more important and has many effects on microalgae bio-growth performances (Wen and Chen, 2001; Xu et al., 2001; Spolaore et al., 2006; Pyle et al., 2008; Hsieh and Wu, 2009; Ho et al., 2011; Morais and Costa, 2014 and Montoya et al., 2014). Physical factors, second category, include pH, hardness, alkalinity, temperature, aeration and the most important is light intensity to the system (Hess and Tolbert, 1967; Ogawa and Aiba, 1981; Martinez et al., 1991; Gonzalez and Bashan, 2000; De-Bashan et al., 2005; Liang et al., 2009; Faramarzi et al., 2010 and Dality et al., 2017).

Microalgae usually use wavelengths between 400-700 nm for photosynthesis. The wavelength microalgae absorb varies depending on the type of microalgae (Blair et al., 2014). Several studies reported that the maximum and minimum number of *Chlorella vulgaris* cells production is in green light (430-465 nm) blue red light (630-665 nm) respectively (Zehnder and Gorham, 1960; Mayo, 1997; Ryu et al., 2009; Al-Qasmi et al., 2012; Khoeyi et al., 2012; Hultberg et al., 2014; Wu, 2015 and Devaraja et al., 2017). Mathy et al., (1996) reported that red light leads to increase of protein and pigments like chlorophyll, astaxanthin, which reflected the positive effect of red light (Wang et al., 2007 and Yek and Chang, 2011). *Chlorella vulgaris* cell size studies revealed that the growth of cells under red light, compared to green light had an increase of approximately 30-60% in diameter (Wang et al., 2007; Yek and Chang, 2011; Blair et al., 2014 and Kim et al., 2014).

Microalgae cultivation systems are divided in open and closed systems. In open systems ambient temperature (> 15°C and < 45°C) is a challenging job. The system depth is another compromise between an adequate light availability for algae cells and enough depth for mixing and avoidance of high ionic fluctuation due to evaporation (Borowitzka, 1999). Most of the open type systems are between 20 and 30 cm deep. Frequently, growth is light limited and typical biomass concentrations range only in between 0.1 and 0.5 g L^{-1} (Scheibenbogen, 1998; Pulz and Borowitzka, 1999; Richmond, 2004; Mata et al., 2010 and Singh and Gu, 2010).

On the other hand, a closed system includes Photo-bioreactors which are made up of transparent material and specifically designed to meet the cultivation requirements of a specific algae. A photo-bioreactor provides a controlled environment, especially with regard to nutritional factors and environmental factors (Olaizola, 2003; Carvalho et al., 2006; Posten, 2009; Morweiser et al., 2010; Anna et al., 2012; Blair et al., 2014; Kula et al., 2014; Asuthkar et al., 2016 and Daliry et al., 2017).

2.11 Role of probiotics in larval rearing

Aquaculture is currently the fastest growing food production sector in the world, but diseases especially bacterial infections remain primary restrictions to its continued development (El-Haroun et al., 2006; Pieters et al., 2008 and Abd El-rhman et al., 2009). It has been well documented by many researchers that the use of antibiotics for a prolonged time period makes drug-resistant micro-organisms and antibiotic residues accumulate in fish flesh and environment (Gildberg et al., 1997; Rollo et al., 2006; El-Haroun et al., 2006; Tukmechi et al., 2007 and Nayak et al., 2007). In addition, antibiotics can affect the normal microflora of digestive tract which is beneficial to host and may be inhibited by treatment (Salminen et al., 1999; Taoka et al., 2006; Panigrahi, 2007; Aly et al., 2008; Mohapathra et al., 2012; Abareethan and Amsath, 2015; Borch et al., 2015; Ibrahem, 2015 and Ma et al., 2015).

In this respect, probiotic bacteria would be useful as food and also serves as biological controllers of fish disease from various pathogens and act as activator for nutrient regeneration (Yasuds and Taga, 1980; Gomez-Gil et al., 2000; Aly et al., 2008 and Kim and Austin, 2008). The most commonly used organism in probiotics is Lactic acid bacteria (LAB) which produces inhibitory compounds like lactic acid, hydrogen peroxide, diacetyl, acetaldehyde and bacteriocin to inhibit harmful microorganism's growth in fish intestine when applied through feed with no harmful effects (Ringo et al., 1998; Ringo and Gatesoupe, 1998; Ringo and Birkbeck, 1999; Gatesoupe, 1999; Nikoskelainen et al., 2001; Salminen et al., 2004; Agrawal, 2005; Abdelhamid et al., 2009; Janczyk et al., 2009; Merrifield et al., 2010; Simon, 2010 and Allameh et al., 2012).

2.12 Effect of probiotic incorporated feed on larval health

The use of probiotics, lactic acid bacteria, as growth promoters in larval rearing of finfishes and shellfishes has been reported by various researchers. Significant increase in the content of crude protein and lipid was recorded in Nile tilapia (*Oreochromis niloticus*), Rohu, (*Labeo rohita*) Rainbow trout (*Oncorhynchus mykiss*), catfishes (*Clarias gariepinus, C. batrachus*), marine shrimp (*Penaeus vannamei, P. monodon, P. japonicas*) when probiotic was supplemented through diet and the also fish weight was boosted from 0.154 g to 6.164 g in Nile Tilapia with improved intestinal health, increase in villous height, improved nutrient absorption, reduction in heterotrophic bacteria in shrimp (Lara-Flores et al., 2003; Abd El-Rhman et al., 2009; Zhou et al., 2010; Pirarat et al., 2011; Jatoba et al., 2011 and Standen et al., 2013). In barbs, the dietary supplementation of lactic acid bacteria resulted high feed conversion ratio, specific growth rate, average daily gain, growth improvement and higher survival rate (Ramasamy, 2003; Dakar and Gohar, 2004; Macey and Coyne, 2005; Kumar et al., 2006; Abdelhamid et al., 2009;

Dohail et al., 2009; Tewary et al., 2011; Lin et al., 2012; Kuhlwein et al., 2013 and Yuniarti et al., 2013).

Intestinal morphological parameters *viz.* villi length, muscular layer thickness, are indicative of healthy gut in fish for digestion and absorption of nutrients. With the increasing lactic acid bacteria to the inclusion level of lactic acid bacteria at 10^7 cfu g^{-1}, intestinal villi of tilapia arranged more regularly with the increased villi length, villi numbers and muscular thickness (Klurfeld, 1999; Deng et al., 2012; Gao et al., 2014 and Zahran et al., 2014).

In case of ornamental fishes, *Poecilia sphenops, Poecilia reticulata, Xiphophorus maculatus, Xiphophorus helleri, Danio rerio,* the incorporation of lactic acid bacteria in diet resulted in growth improvement, elevated specific activities of proteases and amylases in the digestive tract and higher survival (Kumar et al., 2006; Nayak et al., 2007; Ghosh et al., 2008; Abrenouch et al., 2010; Dharmaraj and Dhevendran, 2010; Gioacchini et al., 2010 and Hernandez et al., 2010). In Clownfish, *Amphiprion ocellaris*, a study was performed to explore probiotic addition in diet and the results showed that the weight doubled in both larvae and juveniles with accelerated metamorphosis and significant increase in gene expression of growth factors (Avela et al., 2010 and Chitra and Krishnaveni, 2013).

2.13 Raceway technology for larval rearing

The larval rearing during initial stage is most critical in the cycle of fish seed production in hatcheries and its rearing under controlled condition requires development of specific culture techniques (Olaniyi and Akinbda, 2013). Ornamental fish farming is mainly practiced in outdoor ponds in the tropical countries due to climatic advantages at a very low stocking density (< 300 fish kL^{-1}). In aquaculture production, survival of early stage larvae is still low (<20%) (Fernando and Phang, 1985; Duray et al., 1997 and Toledo et al., 1999).

Compared to other traditional techniques, raceways with recirculation aquaculture systems (RAS) are more advantages in the production of ornamental fish, as such systems give higher survivals, better growth performance with high fish stocking densities due to easier and better management (Mock et al., 1970; Jarboe and Grant, 1996; Sahin, 2001; Kolkovski et al., 2004 and Faulk et al., 2007). Additionally, RAS based raceway technology enable farmers to produce more uniform size fish with increased production in small area through zero water exchange (Vine and Kaiser, 1996 and Kaiser et al., 1998). In this concern, many workers have placed their opinions. Hitzfelder *et al.* (2006) and Faulk *et al.* (2007) reported significant growth in Cobia larvae (from ~4.0 to 14.6 mm) cultured in raceways than in ponds. Another study on swordtail (*Xiphophorus helleri*), showed higher survival and growth rate (Vine and Kaiser 1996; Oliver and Kaiser 1997 and Ako et al., 2005). Similar reports were postulated for marine finfish larvae including cod, silver hake (*Merluccius bilinearis*), gilthead seabream (*Sparus aurata*), yellowtail flounder (*Pleuronectes ferrugineus*) and plaice (*Pleuronectes platessa*) (Buckley et al., 1993; Hagen, 1993; Parra and Yufera, 2000 and Rabe and Brown, 2000).

2.14 Effect of stocking densities on larval rearing

High densities are known to negatively affect fish growth through a variety of channels including general appetite reduction (Wendelaar Bonga, 1997), increased aggression (Hecht et al., 1996), poor water quality (Yu and Perlmutter, 1970) and increased competition for food (Hagen, 1993). On the other hand, optimum stocking density gives higher growth rate, survival, feed conversion ratio, specific growth rate in stipulated area with minimum water.

Gorelsahin et al., (2018) and Oliver and Kaiser (1997) reported very low final mean weight (0.59 g) and survival (63%) in the fishes reared in outdoor ponds compared with in RAS based raceways (1.55 g weight and 99% survival) at optimal stocking density of 3-7 nos. L^{-1} for the guppy (*P. reticulata*) and swordtail (*X. helleri*). Similarly, higher survival and growth rate was observed when Angel fish (*Pterophyllum scalare*), koi (*Cyprinus carpio*), Frenatus (*Epalzeorhynchus frenatus*) and goldfish (*Carassius auratus*) (Degani, 1993; Jha and Barat, 2005; Jha, 2010 and Jahedi et al., 2012) were reared at much lower stocking rates (250–400 fish kL^{-1}) in RAS. Several studies reported that, specific growth rate, survival, average daily gain were inversely related to stocking density during fish rearing (Holm et al., 1990; Haylor, 1991; Bjornsson, 1994; Huang and Chiu, 1997; Irwin et al., 1999; Ma et al., 2006; Aksungur et al., 2007; Hasan et al., 2010 and Tibile et al., 2016).

CHAPTER 3

Materials and Methods

3.1 Induced breeding of *Danio dangila* and *Puntius sophore*

3.1.1 Sample collection and experimental site for animals

Totally 1000 fishes were collected from different locations of North-East Nosa using drag net (180cm × 75 cm, 5mm mesh). *Puntius sophore* (450) was collected from Loktak Lake located near Moirang in Manipur (Lat. 24°33′33.62″ N, Long. 93°48′52.75″ E) (Plate 1) while *Danio dangila* (550) was collected from Brahmaputra stretch near Rajabari Grant in Assam (Lat. 26°28′24.44″ N, Long. 92°32′02.3″ E) (Plate 2). Live fishes were transported to the ornamental fish hatchery situated at ICAR Research complex for North-East hilly region, Manipur (Lat. 27°49′42.67″ N, Long. 93°55′34.45″ E) (Plate 3 and 4) in polythene bags partially filled with oxygen following the method of Rosten (2009) and kept for acclimatization (Plate 5). Species identification was done based on Menon (1954), Talwar and Jhingran (1991), Vishwanath (2010) and Dahanukar (2010).

Danio dangila (Hamilton, 1822) locally called Moustached danio (Angami, 2012) have brilliant coloration of body with brownish to olive on back and silvery yellowish in belly and flanks. Three to four silvery bluish lines on flanks and the bands of the body break up anteriorly forming a mottled pattern. Two well-developed pairs of barbels (rostral and maxillary) were observed. Maxillary pairs were slightly longer than the rostral pairs. Upper angle of gill opening occasionally shows a dusky spot. Lateral line was complete with 36-40 small to cycloid scales and 12 to 14 pre-dorsal scales (Vishwanath, 2010) (Plate 6).

Plate 1: Loktak Lake, Moirang, Manipur

Plate 2: Brahmaputra stretch, Rajabari, Assam

Plate 3: Fish sample collection

Plate 4: Ornamental fish hatchery, ICAR sub-center for NEH region, Manipur

Plate 5: Fish packing at Manipur

Puntius sophore (Hamilton, 1822) locally called Spotfin swamp barb (Mahapatra, 2010) is a very plentiful shoaling fish, upperciably smaller, beautiful with silver body and back is grey-green to brownish, flanks with bluish luster, a deep black round blotch at the base of the caudal fin and on central part of dorsal fin, a long red line on anterior part of the body make fish as very beautiful animal in aquarium keeping. Barbles were absent. Dark spot was present on dorsal fin in between 4th & 8th fin rays. Another black spot was present on caudal fin base in between 21st & 23rd scales. Lateral line was incomplete with 26 - 28 scales. Scales at the base of dorsal fin were five while four scales were observed at anal fin base. This species matures at 4-6 cm but can attain a maximum length of about 13 cm (Dahanukar, 2010) (Plate 7).

Thereafter, the fishes were repacked in oxygenated polythene bags and brought to Aquatic Rainbow Technology Park (ARTP), Dr. MGR Fisheries College & Research Institute, TNJFU Madhavaram campus, Chennai (Lat. 13° 9′42.81″N, Long. 80°15′1.86″E) through air transport following same method described earlier (Plate 8). The fishes were stocked at the rate of 2 nos. L^{-1} in circular fiberglass reinforced plastic (FRP) tanks of 2000L capacity attached with RAS facility for a period of 2 months (Plate 9 and 10). The fishes were daily fed with live feed such as *Moina micrura, Thermocyclops hyalinus, Tubifex tubifex, Eudrilus Eugeniae* and supplementary feed having 40% crude protein at the rate of 4-5% of body weight. Fishes were observed weekly for morphological indicators of maturity and health issues.

Plate 6: Moustached danio, *Danio dangila*

Plate 7: Sophore barb, *Puntius sophore*

Plate 8: Multi-species ornamental fish hatchery, TNJFU, Madhavaram

Plate 9: Acclimatization of collected ornamental fishes

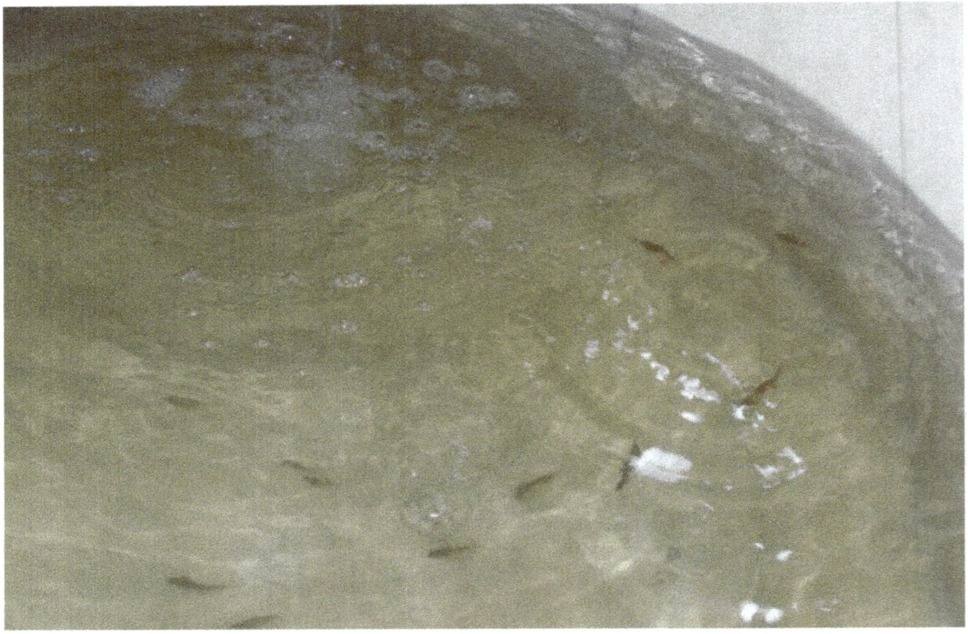

Plate 10: Rearing of collected wild stock in RAS - FRP tanks

3.1.2 Physico-Chemical parameters of water

Physico-chemical parameters are of very much important for the rearing and breeding of *Danio dangila* and *Puntius sophore*. Reverse Osmosis (R.O.) and U.V. filtered water after maturation (24 hrs aeration) was used for domestication and breeding of the fishes. In all the rearing and breeding experimental tanks, water quality parameters were maintained at optimum level *viz.* pH - 6.9-7.2; Total Dissolved Solids - 30-50 mg L^{-1}; Dissolved Oxygen - 5.5-6.5 mg L^{-1}; Free CO_2 - <1.0 mg L^{-1}; Alkalinity - 45-50 mg L^{-1} and were determined weekly by Standard Methods (APHA, 2012). In the rearing tanks, temperature was maintained within the range of 21-28°C with the help of thermostat, regulated water heaters.

3.1.3 Selection of brooders

After rearing and acclimatization of wild fishes, 120 matured females (60 each of *D. dangila* and *P. sophore*) and 240 matured males (120 each of *D. dangila* and *P. sophore*) in a sex ratio of 1:2 (male : female) were selected for experiment. Matured animals were segregated based on their morphological characters. Morphological characters of *Danio dangila* has been summarized in Table 2 (Angami, 2012; Langer, 2013) whereas for *Puntius sophore* in Table 3 (Menon 1999; Dahanukar, 2010). Mean weight of *D. dangila* males was 5.2±0.35 g and females weighed around 6.50±0.23 g. Mean weight of *P. sophore* males was 4.25±0.24 g and of females it was 6.14±0.35 g. For further confirmation of maturity in females, oocyte diameter was measured by obtaining *in-vivo* biopsy of ovary using a polyethylene cannula of 2 mm diameter and observed under trinocular microscope attached with camera showed ova diameter of 1.05-1.38 mm for *D. dangila* and 0.95-1.15 mm ova diameter for *P. sophore* (NLCD-120E Lawrence & Mayo make) (Shehadeh et al., 1973).

Table 2: Morphological characters of broodstock in *Danio dangila*

Character	Male	Female
Size	Smaller	Larger
Body colour	Bright	Slightly dull
Belly shape	Slender and ventrally straight	Bulgy and ventral curved
Fin texture (Anal)	Slightly rough with red linings	Soft with whitish yellow linings
Anal Opening	Oval	Round

(Menon, 1999 and Dahanukar, 2010) (Plate 11 A and B).

Table 3: Morphological characters of broodstock in *Puntius sophore*

Character	Male	Female
Size	Smaller	Larger
Body colour	Bright (Prominent pink spot on gill)	Slightly dull (Faint pink spot)
Belly shape	Slender and straight	Bulgy and ventral curved to anal fin posture
Fin texture (Anal)	Slightly rough and brick red in color	Soft with whitish yellow to hyaline in color
Anal Opening	Oval	Round

(Angami, 2012 and Langer, 2013) (Plate 12 A and B).

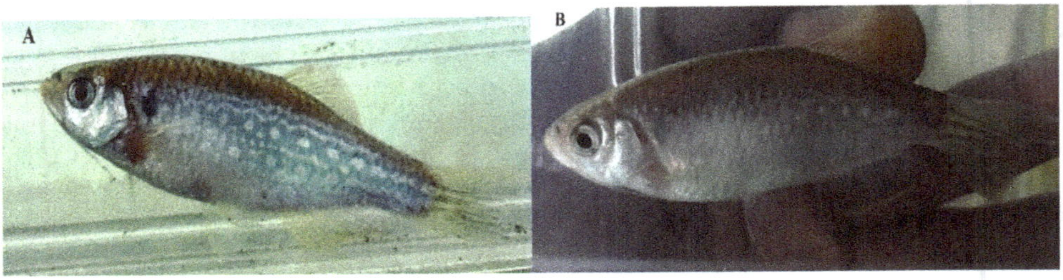

Plate 11: *Danio dangila* broodstock A. Male and B. Female

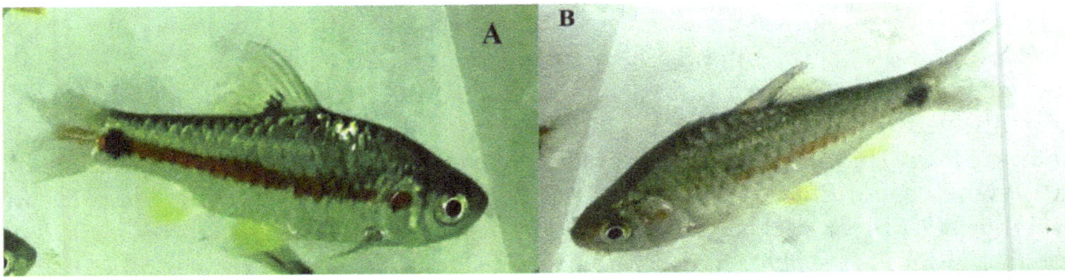

Plate 12: *Puntius sophore* broodstock A. Male and B. Female

3.1.4 Setting of spawning tank

Totally 20 aquarium tanks (60 cm × 45 cm × 30 cm; L×H×W) were filled with R.O. and UV filtered water upto 30 cm height provided with biological filter and round the clock aeration. Hapa (30 cm × 30 cm × 30 cm) made up of velon screen (10 mm mesh) were fixed in each tank and the tank water was allowed to mature for 24 hours (Plate 13).

3.1.5 Procurement of hormone and preparation

Synthetic hormones *viz.* WOVA-FH, HCG and Ovatide were obtained from Biostadt Nosa Ltd., Worli, Mumbai, India. A syringe of 1 mL having 40 divisions was selected for convenience in which each unit measures 0.025 mL of hormone. Hormone was diluted in equal quantity of distilled water which made hormone concentration at 0.0125 mL unit[1] suitable for 5-7 g sized of fishes (Mangesh et al., 2017).

3.1.6 Induced breeding

After settings of breeding tanks, hormones at different dosages were used to induce fishes (Table 4). Matured females were injected twice giving 6 hours gap while males were injected once during the resolving dose of female at the base of pectoral fin (intra-peritonially) (Plate 14). While injecting care was taken to reduce the chance of cardiac puncture by needle. The present experiment followed Completely Randomized Design (CRD) using four treatments in addition to a control performed in triplicates for each hormone. After hormonal inducement, reproductive behavior of fishes was observed periodically.

Plate 13: Experimental setup for induced breeding of selected ornamental fishes

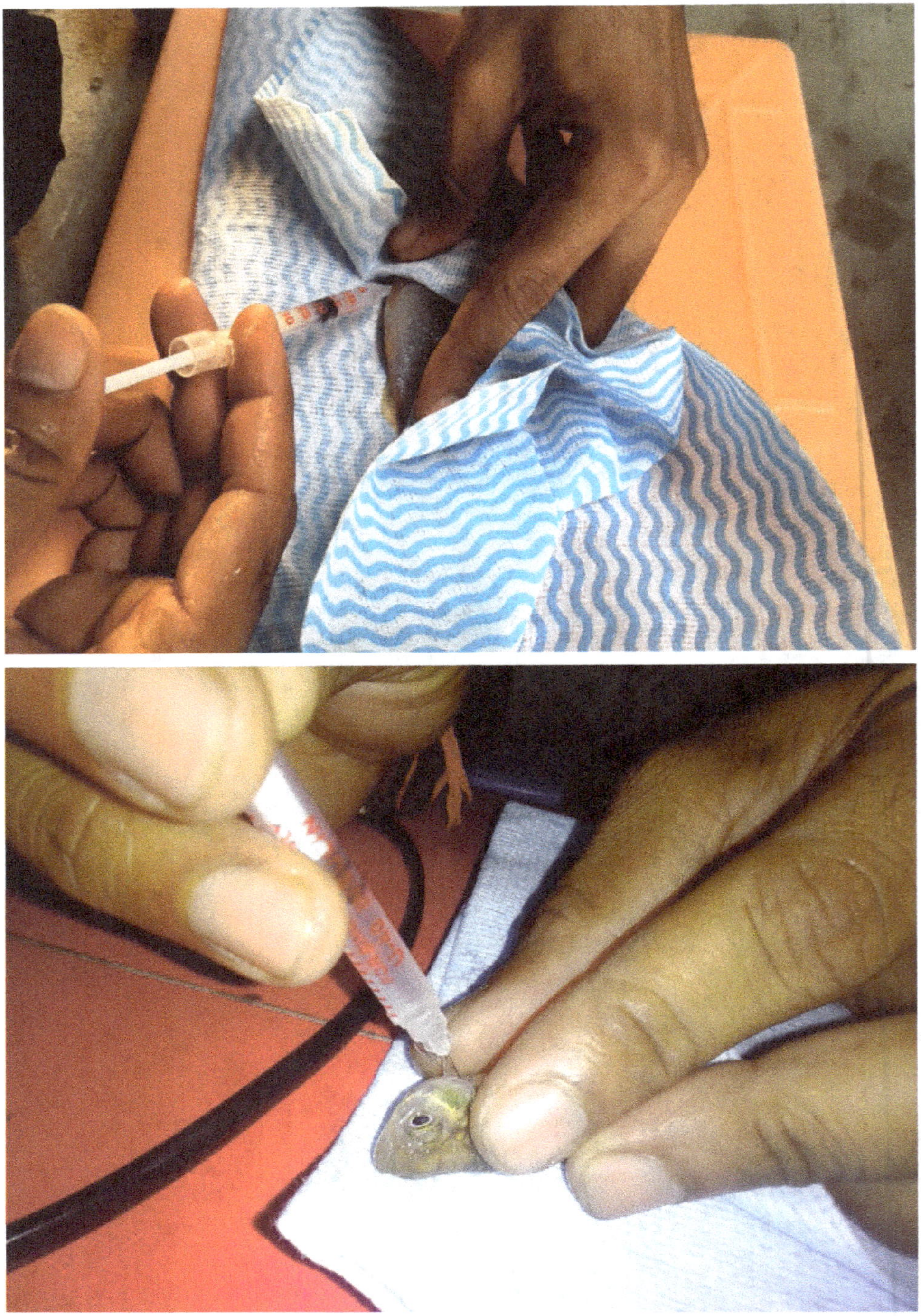

Plate 14: Intra-Peritonial injection to *Danio dangila* and *Puntius sophore*

Table 4: Treatments based on hormonal dose applied to induce fishes

Treatment	Hormone dosage Female	Hormone dosage Male
T_0	No inducement	No inducement
T_1	0.3 mL kg^{-1}	0.1 mL kg^{-1}
T_2	0.4 mL kg^{-1}	0.2 mL kg^{-1}
T_3	0.5 mL kg^{-1}	0.3 mL kg^{-1}
T_4	0.6 mL kg^{-1}	0.4 mL kg^{-1}

3.1.7 Breeding performance

Breeding performance of individual female fishes were evaluated based on the following criteria *viz.* breeding behavior, latency period (hours), number of eggs spawned, fertilization rate (%) and hatching rate (%). The latency period is the time gap between hormonal injection and the first appearance of spawned eggs (Kumar et al., 2018). Number of eggs spawned from each female after spawning was determined through random sampling of released eggs in a 5 mL graduated measuring cylinder. The total number of eggs in 0.5 mL was counted in triplicate and multiplied with total volume of eggs released and calculated by following mathematical expression (Behera et al., 2010) (Plate 15 and 16).

Number of eggs spawned

= Total number of eggs in 0.5 mL ×Total volume of eggs (mL)

(Behera et al., 2010)

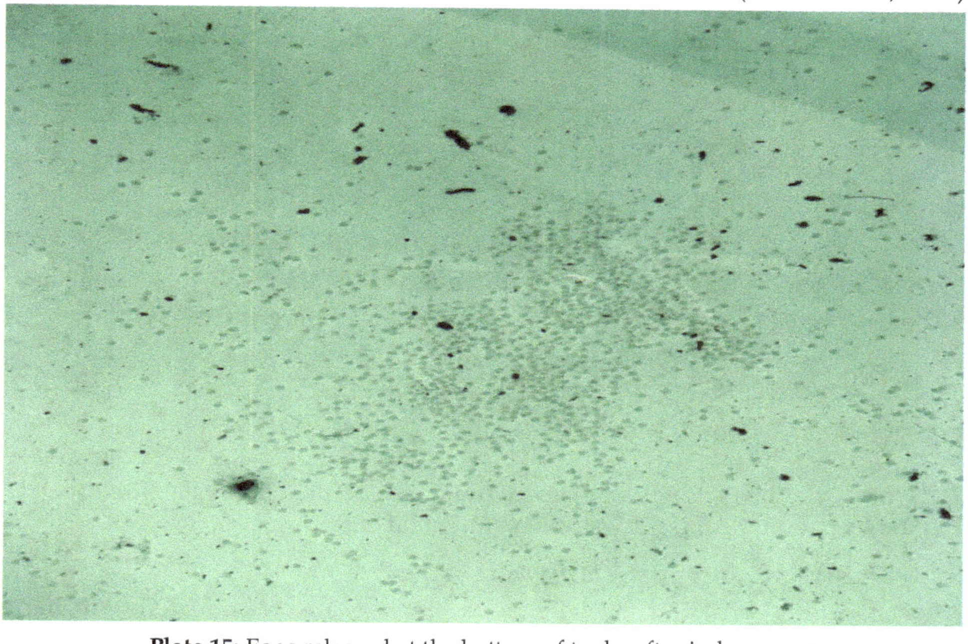

Plate 15: Eggs released at the bottom of tanks after inducement

Plate 16: Counting of fertilized eggs

The fertilization rate of eggs was determined by randomly taking samples of approximately 100 eggs from the total eggs in a glass petri-dish (Pronob Das et al., 2016). Fertilized eggs having intact nucleus were only considered for calculating percentage of fertilization and is calculated by dividing fertilized eggs with total number of eggs observed multiplied by hundred.

$$\boldsymbol{Fertilization\ Rate}(\%) = \frac{Number\ of\ Fertilized\ Eggs}{Total\ Number\ of\ Eggs\ in\ Sample} \times 100$$

(Pronob Das et al., 2016)

Hatching rate was calculated by dividing hatchlings with total number of fertilized eggs multiplied by hundred.

$$\boldsymbol{Hatching\ Rate}(\%) = \frac{Total\ Number\ of\ Spawns}{Total\ Number\ of\ Fertilized\ Eggs} \times 100$$

(Pronob Das et al., 2016)

One day old hatchlings were transferred and maintained in 200 litre capacity FRP tanks having RAS facility stocked at the rate of 50 nos. L^{-1}.

3.1.8 Embryonic and post-embryonic stages

After spawning, brooders from the aquarium tanks were removed with hapa and transferred to brood stock tanks (FRP) allowing eggs in aquarium tanks for incubation and hatching. The eggs were collected using dropper for observing embryonic developments and measurements were done under trinocular microscope at 40X (NLCD-120E, Lawrence & Mayo) while post-embryonic developments were observed under stereozoom microscope attached with camera at 10X (EM 33 MEIJI).

3.2 Mass production of fresh water Rotifers (*Brachionus calyciflorus*) in outdoor culture system

3.2.1 Procurement and preparation of Rotifer (*Brachionus calyciflorus*) stock

Fresh water Rotifer, *B. calyciflorus* (phylum Rotifera) was obtained from the wild and sifted through 600, 212, 125, 75 and 38 μm sieves arranged one above another in a decreasing downward order and observed under light microscope at 10X magnification for confirmation of desired fauna (NLCD-120E Lawrence & Mayo). The morphometric identification was done based on following characters. In healthy rotifer, lorica saccate was soft and ovoid shaped with a smooth surface. The antero-dorsal margin was with four spines (two on each side) on a U-shaped sinus. All spines were triangular with a wide base and sharp tips. The anterior ventral margin was smooth with a medial sinus. Three antennae were present. One in the U-shaped sinus in between antero-dorsal spines and two were on lateral side of the lorica. Trophi was malleate type with short and hollow fulcrum. Ramus was asymmetrical and Uncus was with four teeth decreasing in size from the ventral to dorsal while sub-uncus was brush like. Manubria were triangular in shape and rounded at the external sickle-shaped margin with flattened and slightly bended at their distal end (Michaloudi et al., 2018).

The stock culture was maintained in 500 mL Erlenmeyer's flasks and fed with Baker's yeast. Sufficient aeration was provided through the perforated stones. Successful starter culture was maximized using batch culture to meet the requirement for experimental trials.

3.2.2 Procurement and preparation of *Chlorella vulgaris, Spirulina major* and *Scenedismus obliquus* cultures

Samples of green water with varieties of microscopic algae were collected from culture ponds constructed at Advance Research Farm Facility (ARFF). Isolation of *C. vulgaris, S. major* and *S. obliquus* from green water sample was achieved first by picking up desired flora followed by agar plating methods (1% bacteriological grade agar) under 3000 lux fluorescent light at temperature 25±1.0°C in incubator (Pringsheim, 1946). Algal colonies were picked up selectively and inoculated into 20 mL test tubes containing same agar media followed by 1000 mL Erlenmeyer flasks (Plate 17 and 18). After getting pure strains, the species were cultured in 1.0 L water in sterilized plastic bottles under light and aeration (Plate 19). Mass culture was performed in cement cistern tanks (1000 L) with light and aeration facility (Plate 20). Water fertilization was done using ground nut oil cake (25%), urea (1%) and single super phosphate (0.5%) and inoculation was added on the same day. Baker's yeast was procured from local market.

3.2.3 Experimental setup

Experiment was done following Completely Randomized Design (CRD) with three trials in 1000 L capacity cement cistern tanks (1×1 meter; D×H) at Advanced Research Farm Facility (ARFF), Dr. MGR FC & RI, TNJFU Madhavaram campus. Rotifers were inoculated at a density of 10 nos. mL^{-1} in tanks. Four feed types (*Chlorella vulgaris, Spirulina major, Scenedismus obliquus* and Baker's Yeast) were extensively used to feed rotifers with continuous aeration (Table 5; Plate 21 and 22).

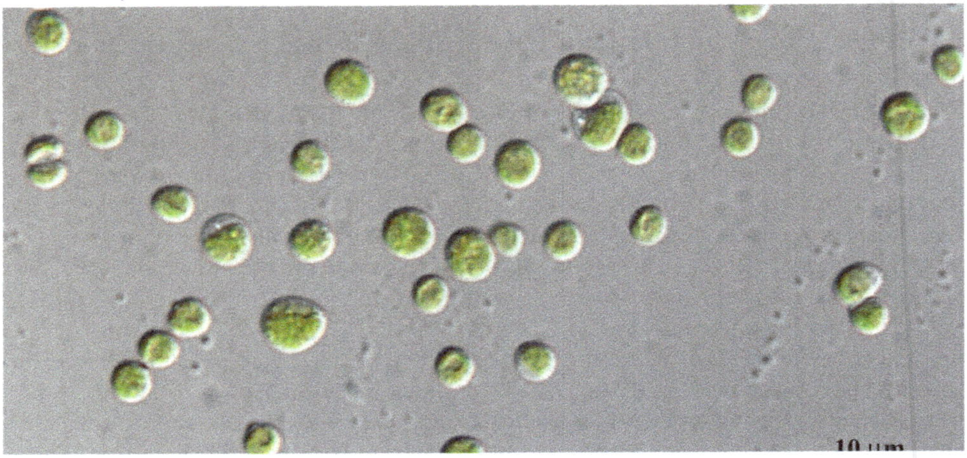

Plate 17: Observation of *Chlorella vulgaris* under microscope (40X)

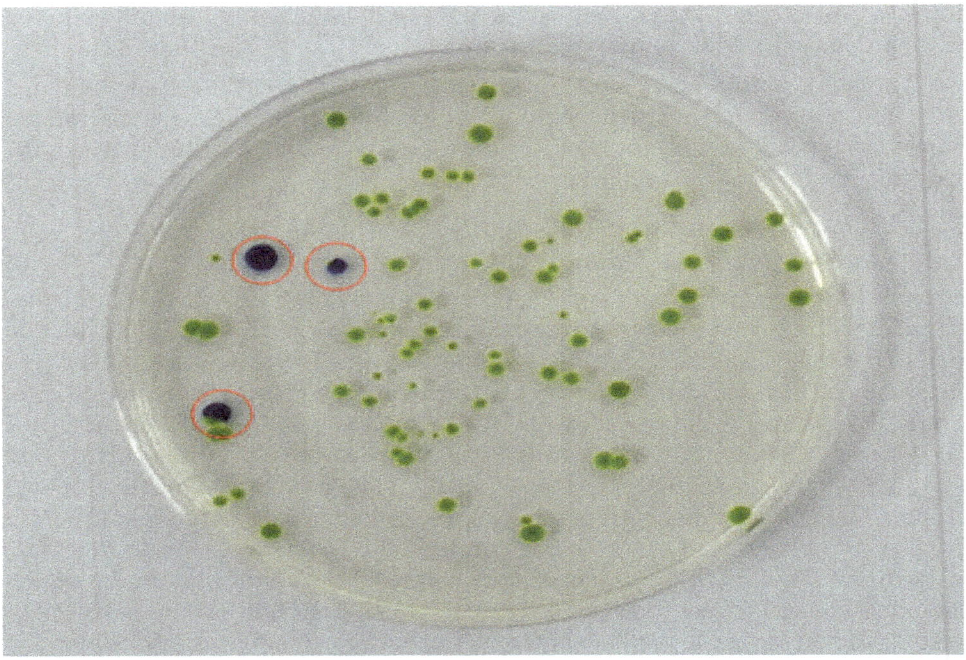

Plate 18: *Chlorella vulgaris* colonies on agarose

Plate 19: Stock culture of *Chlorella vulgaris*

Plate 20: Mass culture of *Chlorella vulgaris*

3.2.4 Sampling of Rotifers

Rotifers were sampled at every 2nd day to determine rotifer density in the morning to minimize stress. A micropipette of 1.0 mL was used to collect sample and rotifer count was observed under trinocular microscope (NLCD-120E Lawrence & Mayo) using Sedgwick Rafter cell counter (S50, PYSER, SGI) (Plate 23). Water quality parameters were determined on every second day by following Standard Methods (APHA, 2012) (Plate 24).

Table 5: Treatments based on different feed type used in culture of freshwater rotifer in outdoor culture system

Treatment	Food types
T_c	No food applied
T_1	Spirulina major
T_2	Scenedismus obliquus
T_3	Chlorella vulgaris
T_4	Baker's Yeast

3.2.5 Feeding intensity

The feed type with highest population growth was used for further trials on extent of feed concentration (2× and 3× day^{-1}) required for obtaining maximum growth of rotifers. Tanks (1000 L) were inoculated at 10 nos. mL^{-1} of rotifer and extensively fed with algae at 2× and 3× day^{-1} intensities in triplicate following Completely Randomized Design (CRD).

Plate 21: Live feed production unit (outdoor)

Plate 22: Experimental setup for outdoor rotifer production

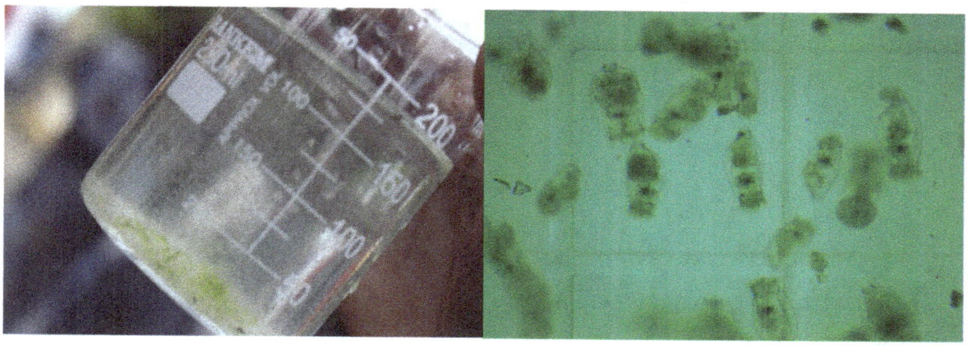

Plate 23: Rotifer sample collection and observation

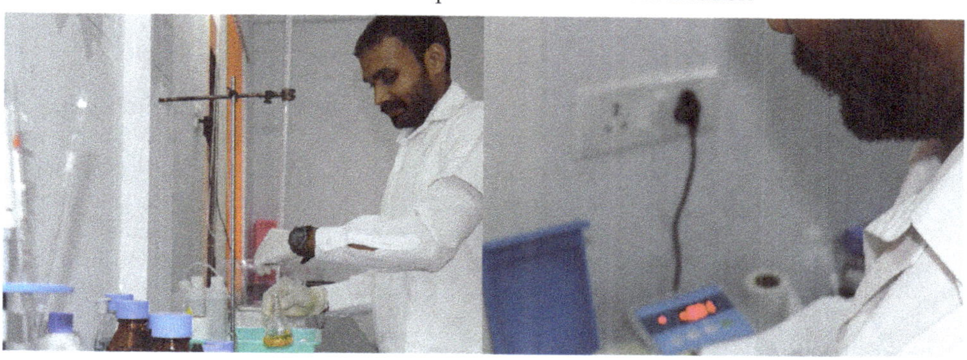

Plate 24: Water quality analysis

3.3 Mass production of *Chlorella vulgaris*

Experiment was conducted at live feed production unit following Completely Randomized Design (CRD) in 500 L capacity cylindrical Fibreglass Reinforced Plastic (FRP) tanks (1.25×0.75 meter; D×H), in triplicate, with continuous aeration and light facility. LED strips of four different light wavelengths *viz.* red, blue, white and green (5 kLux capacity) were procured from local market (Table 6; Plate 25).

Table 6: Treatments based on light wavelengths for biomass production and pigment enhancement of *Chlorella vulgaris* in internal photobioreactor

Treatment	Light wavelength	Wavelength (nm)
T_c	White	380 to 750
T_1	Red	620–750
T_2	Green	495–570
T_3	Blue	450–495

3.3.1 Procurement of *C. vulgaris* and media for culture

The stock culture of *C. vulgaris* was cultured in 1.0 L water in sterilized plastic bottles under light and aeration. Mass culture was performed in 500 L capacity cylindrical FRP tanks. Water fertilization was done using ground nut oil cake (25%), urea (1%) and single super phosphate (0.5%) and inoculation was added on the same day.

Plate 25 Live feed production unit (FRP based)

3.3.2 Design of internal photobioreactor

Commonly used aeration system was modified by making 1 mm holes in polyvinyl chloride (PVC) pipes for aeration purpose operated by 1 HP air blower (CLEANTEK). Inside the tank, underwater light facility was developed by inserting Light Emitting Diodes (LED) of 5 kLux capacity (blue, red, white and green) in a waterproof transparent acrylic pipe of 120 cm length and 1.27 cm diameter. PVC pipes for aeration were placed horizontally at the bottom of tanks while acrylic pipes with LED were placed vertically in water column (Plate 26).

3.3.3 Control of Ciliates

Ciliates have been a major problem for algae culture. They have very high ingestion, grazing, growth rate and result in total destruction of dense cultures of algae within 48-72 h. Ciliate affected algal culture tanks (1000 L) were applied with salt at different dosage (30g L^{-1}, 60g L^{-1}, 90g L^{-1} and 120 g L^{-1}) to control ciliates growth in triplicate other than chemicals and ozone sterilization (Table 7).

Table 7: Effect of common purified salt on control of ciliates in algae, *C. vulgaris*, **culture tanks**

Treatment	T_c	T_1	T_2	T_3	T_4
Common purified salt (g L^{-1})	0	30	60	90	120

3.3.4 Measurement of *Chlorella vulgaris* biomass and specific growth rate

Algal cell count was estimated using Neubauer haemocytometer (Bauer, 1990). Ruled areas were centrally covered with cover slip and algal sample was placed in the "V" groove of the metal surface. Chambers were filled gradually for even distribution of algal cells without air bubbles. Counting was done from the top left square and only those cells were counted, which lay within or on the boundary line of the square under trinocular microscope (Lawrence & Mayo, NLCD-120E). All the cells counted in individual block were added up and cell density was calculated by using the following mathematical expression:

$$\boldsymbol{D}\left(\boldsymbol{Cells\ mL^{-1}}\right) = \frac{\left(Total\ count \times 10^{4}\right)}{Blocks}$$

(Hewette et al., 1991)

The specific growth rate (μ) at the exponential phase was calculated using the following equation:

$$Specific\ Growth\ Rate\ (\mathbf{\imath}) = Ln\frac{(X_2 / X_1)}{(t_2 - t_1)}$$

(Hewette et al., 1991)

Where, X_2 and X_1 are the final and initial biomass concentrations at times t_2 and t_1 respectively.

3.3.5 Determination of protein content in *C. vulgaris*

Lowry (1951) method was used to analyze total protein content of *C. vulgaris*. Based on this method, Deoxycholate (DOC) solution (0.1 mL) was added in 1 mL of sample and kept at room temperature for 10 min. Again, Trichloroacetic acid (TCA) (0.1 mL) was added in the same sample and which was centrifuged at 7500 rpm for 10 min. After removal of the supernatant, Lowry solution (1 mL) was added in the precipitate and allowed to stand at room temperature for 20 min. Folin reagent (1 mL) was then added in the sample and kept for 30 min. Finally, the absorbance of samples was measured at 750 nm using visible spectrophotometer (Deepvision, 2305) and the results were evaluated based on the standard curve (Lowry et al., 1951).

Plate 26: Layout and set up of internal photobioreactor with LEDs in culture tanks of *C. vulgaris*

3.3.6 Determination of lipid content in *C. vulgaris*

Total lipid content was determined following methodology given by Bligh and Dyer (1959) with slight modifications. Cultured *C. vulgaris* was isolated and freeze dried at -20°C (Everflow Scientific). Freeze dried algae sample (0.05g) was used for lipid extraction in a mixture of 2 ml chloroform and 2 ml methanol (1:1). Initially, NaCl (2 ml of 0.88%) was added to the mixture to improve the performance of lipid extraction. The mixtures were shaken vigorously and centrifuged at 5000 rpm for 3 minutes. Methanol or water (top phase) and chloroform (bottom phase) were observed. The chloroform layer was further purified using sodium sulphate anhydrous powder and collected into the weighted beaker. The organic solvents inside were allowed to evaporate under nitrogen gas. The remaining lipids were weighed which provided the percentage of lipids in algal dry weight. Lipid productivity (mg mL^{-1}) was calculated using the mathematical expression:

$$\text{Lipid (mg mL}^{-1}) = \text{Lipid content (\%)} \times \text{Dry biomass (mg L}^{-1})$$

(Bligh and Dyer, 1959)

3.3.7 Determination of pigment content

The amount of chlorophyll (a & b) was measured using the method of Strickland and Parsons (1972). Algal sample (5 mL) was filtered through GF/C filter paper to determine the concentration of the pigments. Each filter paper was placed in 90% acetone at 4°C in dark for 24 hours and centrifuged at 3500 rpm for 5 min. Chlorophyll a & b were analyzed by measuring the absorbance at 630, 645, 665 and 750 nm using a visible spectrophotometer (Deepvision, 2305). The amount of astaxanthin was calculated using the method of Boussiba and Vonshak (1991). Initially, algal sample (5 mL) was centrifuged at 4000 rpm for 10 min to remove the supernatant followed by adding of 30% methanol followed by adding 5% potassium hydroxide into the precipitate and kept at 65°C for 15 min to ensure removal of the chlorophyll. The supernatant was decanted followed by three times washing with distilled water. Then, dimethyl sulphoxide (DMSO) (5 mL) was added over the precipitate and sonicated for 10 min (3.0 kHz). The samples were analyzed by calculating the amount of astaxanthin at 490 nm using a visible spectrophotometer (Deepvision, 2305) (Davis, 1976).

3.4 Mass production of freshwater Rotifers (*Brachionus calyciflorus*) in indoor culture system

3.4.1 Sourcing of Rotifers (*Brachionus calyciflorus*)

Fresh water Rotifer, *B. calyciflorus,* (L-type) of 350-500 μ lorica length having eggs and algal species, (*C. vulgaris, S. major, S. obliquus*) were procured from Live Feed Unit, Advance Research Farm Facility (ARFF), TNJFU, Madhavaram campus. Culture of algal species was performed in 4000 L (3.5×1×1.2 meter; L×B×H) capacity rectangular tanks after fertilization with ground nut oil cake, urea and single super phosphate at 25%, 1%, 0.5% respectively using under water light (5000 lux) and aeration facility (Plate 27).

3.4.2 Experimental design

Experiment was done following Completely Randomized Design with three trials in 5000 L capacity tanks (4×1×1.2 meter; L×B×H) attached to Recirculatory Aquaculture System (RAS). Four feed types (*Chlorella vulgaris, Spirulina major, Scenedismus obliquus* and Baker's Yeast), algal consumption rate and daily harvest rates, at the rate of 10%, 30%, 50% and 70%, were observed for mass production and for maintaining culture at its stationary phase throughout the season (Plate 28).

Rotifers were inoculated at a density of 10-12 nos. mL^{-1} in 5000 L (4.5×1×1.2 meter; L×B×H) capacity rectangular tanks and extensively fed with different algae cultured using submersible pump. The culture water pH, temperature, hardness and total ammonia nitrogen level was maintained constant at 7.2±0.8, 28±4°C, 250±40 ppm and ≤0.5 ppm respectively.

Table 8: Treatments based on different feed type used in the culture of freshwater rotifer in indoor culture system

Treatment	Food type
T_c	No food applied
T_1	Spirulina major
T_2	Scenedismus obliquus
T_3	Chlorella vulgaris
T_4	Baker's Yeast

3.4.3 Sampling of Rotifers

Rotifer density in each tank was estimated by taking three random samples, in the morning between 8.00 to 10.00 AM on every 2nd day using micropipette (1 mL).

Live rotifers after fixation with two drops of ethanol were observed in a Sedgewick Rafter Counting Chamber (S50, PYSER, SGI), under a trinocular microscope (NLCD-120E Lawrence & Mayo). The rotifers with empty and transparent lorica were not counted (Plate 29).

Plate 27: RAS based live food production unit

Plate 28: Experimental setup for live food and algae culture

3.4.4 Estimation of algal ingestion rate by rotifer

The rotifers (200±1.12 nos.) were housed in 500 mL fiberglass jars provided with varying quantities of algae (10×10[6], 20×10[6], 40×10[6] and ×10[6] cells mL^{-1}) in triplicates, while one was served as control without food for 24 hours (Table 9). Algal cell ingestion rate was estimated using Neubauer haemocytometer (Bauer, 1990). Ruled areas were centrally covered with cover slip and algal sample was placed in the "V" groove of the metal surface and filled without air bubbles. Counting was done from the top left square and only those cells were counted, which lay within or on the boundary line of the square. All the cells counted in individual block were added up and cell density was calculated by the mathematical expression:

$$\boldsymbol{D\left(Cells\ mL^{-1}\right)} = \frac{\left(Total\ count \times 10^{4}\right)}{\text{Blocks}} \qquad \text{(Hewette et al., 1991)}$$

Ingestion rates on the basis of filtration rate of rotifer were calculated following Yufera and Pascual (1985) mathematical expression:

$$\boldsymbol{Ingestion\ Rate(IR)} = F \times \sqrt{C_0} \times C_t \qquad \text{(Hewette et al., 1991)}$$

$$\boldsymbol{Filtration\ Rate(F)} = \frac{Ln C_0 - L_n C_t}{V \times t} \qquad \text{(Hewette et al., 1991)}$$

Where C_0 and C_t are initial and final algal cell densities, t = time, F = Filtration rate and V = rotifer density.

Table 9: Treatments based on algal cell concentration on ingestion and filtration rate of freshwater rotifer, *Brachionus calyciflorus*

Treatments	T_c	T_1	T_2	T_3	T_4
Algal cells mL^{-1}	00	60×10^{6}	40×10^{6}	20×10^{6}	10×10^{6}

3.4.5 Mass culture and harvesting protocol

The present experiment was conducted in 5000 L capacity tanks to standardize harvesting rate for maintaining the rotifer culture at its stationary phase throughout the season. After rotifer inoculation (10 nos. mL^{-1}), *C. vulgaris* was extensively used to feed rotifers at the rate of 2.3×10^{5} cells mL^{-1} initially and was gradually increased to meet the food requirement. Before harvesting rotifer, count was observed at 2nd, 4th, 6th and 8th days by tanking three random samples from each tank using 1 mL micropipette in a Sedgewick Rafter Counting Chamber. Tanks were harvested daily at the rate of 10%, 30%, 50% and 70% of the total volume (Table 10) and checked quantum of harvest using mathematical expression:

$$\textbf{Total harvest} = \textit{Rotifers mL}^{-1} \times \textit{Quantity of water harvested (mL)}$$

The harvested tanks were again refilled with same quantum of algal water using submersible pump while algal tanks were refilled with filtered water. Daily observation for proper aeration and sufficient light supply in the algal culture tanks, deposition of culture media, and accumulation of algae on tank walls were strictly monitored. Daily water quality was observed for water color, smell and water insect contamination. All trials were conducted in the indoor RAS based live food production unit.

Table 10: Effects of different harvesting rates on rotifer production in RAS based system

Treatments	T_c	T_1	T_2	T_3	T_4
Harvest rate (%)	00	10	30	50	70

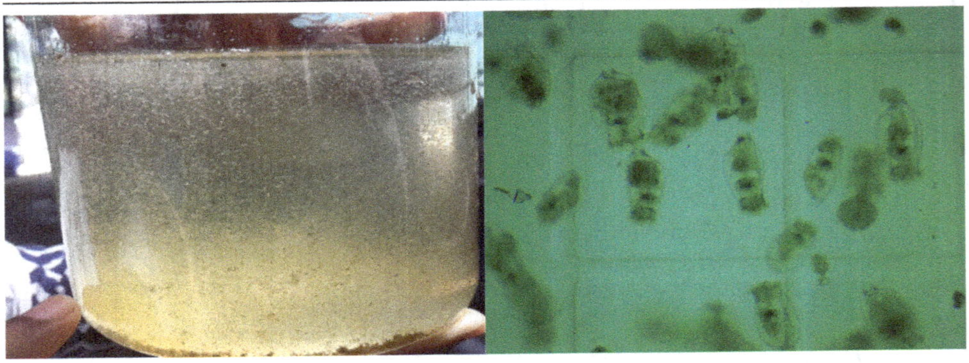

Plate 29: Rotifer sample collection and observation

3.4.6 Maintenance of Algal culture

Algal culture tanks were monitored periodically for water quality parameters, maximum cell count, deposition of algae on side walls, dead algae, proper aeration and light and deficiency of required nutrients. To fulfill the nutrient requirements, algal tanks were fertilized after every two weeks for preventing death of algae (Plate 30).

3.4.7 RAS (Recirculatory Aquaculture System) facility for continuous culture of rotifer

The RAS facility for continuous culture system consists of four components; algal culture, rotifer culture, filtration facility and harvesting units. Algal culture tanks were facilitated with under water aeration and light facilities. Commonly used aeration system was modified by making 1 mm holes in polyvinyl chloride (PVC) pipes for aeration purpose operated by 2 HP air blower (CLEANTEK) and under water light facility was developed by inserting green light emitting diodes (LED) of 5000 lux capacity in a water-proof transparent acrylic pipe of 200 cm length and 1.27 cm diameter. PVC pipes for aeration were placed at the bottom of the tanks while acrylic pipes with LED were placed horizontally in water column. Rotifer culture tanks were provided with under water aeration facility following same method with separate inlet and outlet valve without light. The filtration facility consists of one sand filter assembled with a layer of activated carbon and sand (500 μ), four wound type cartridge filters (10 μ) of 50.8 cm diameter arranged in series and one UV filter with three lamps of 30 watts assembled in a stainless steel body (15.24 cm diameter) of 5 m^3 h^{-1} capacity. Harvesting facility was provided with separate valves for each tank which opens in a harvesting pit. After collection of rotifers through plankton net (20 μ mesh), the water moves to 5 m^3 capacity sump constructed near culture tanks through underground PVC pipes by gravity. The water collected in sump was re-utilized for algal culture after filtration with the help of centrifugal pump (2 HP). The design of RAS system and filters for live food production has been illustrated in Plate 31, 32, 33 and 34.

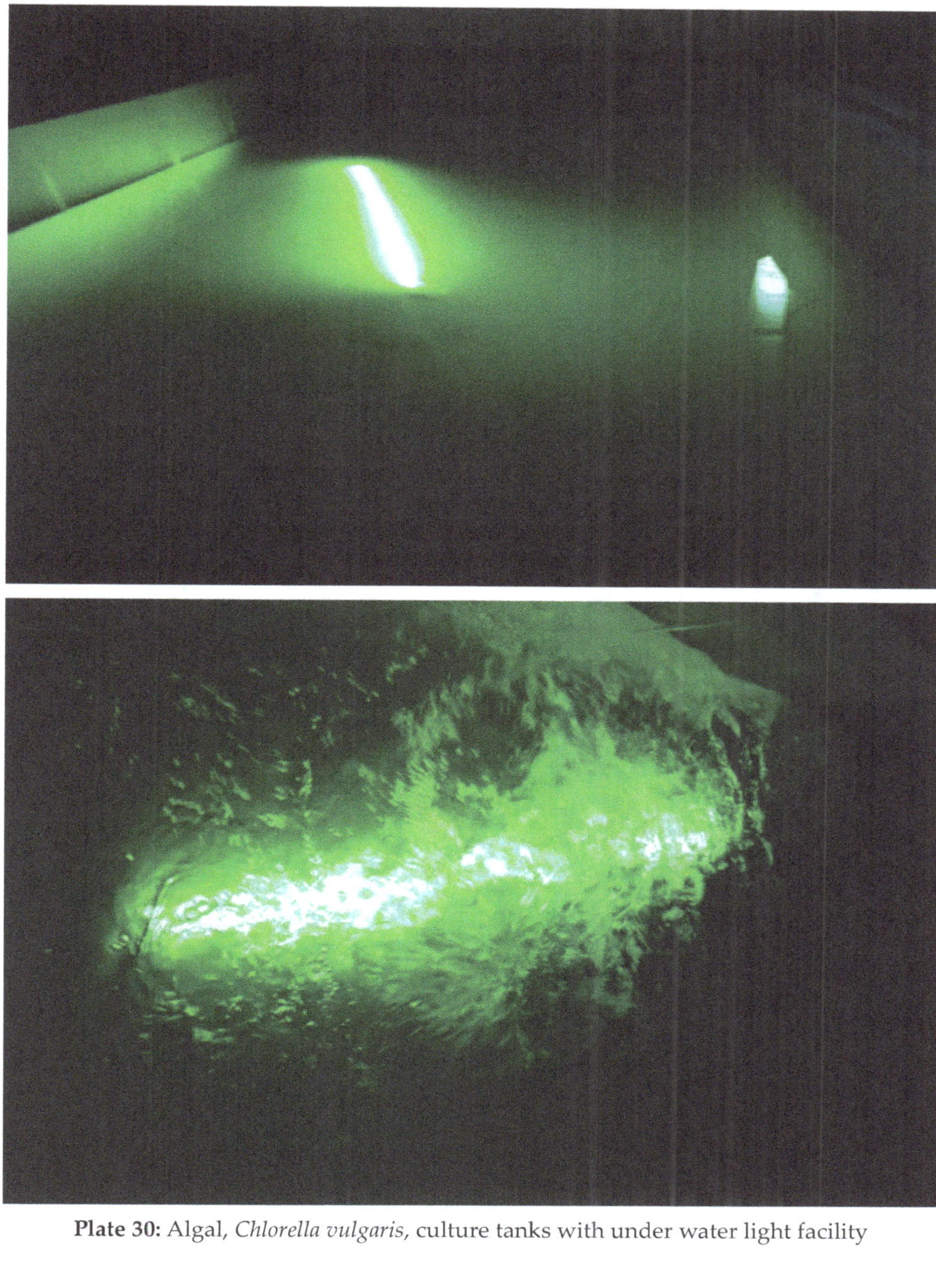

Plate 30: Algal, *Chlorella vulgaris,* culture tanks with under water light facility

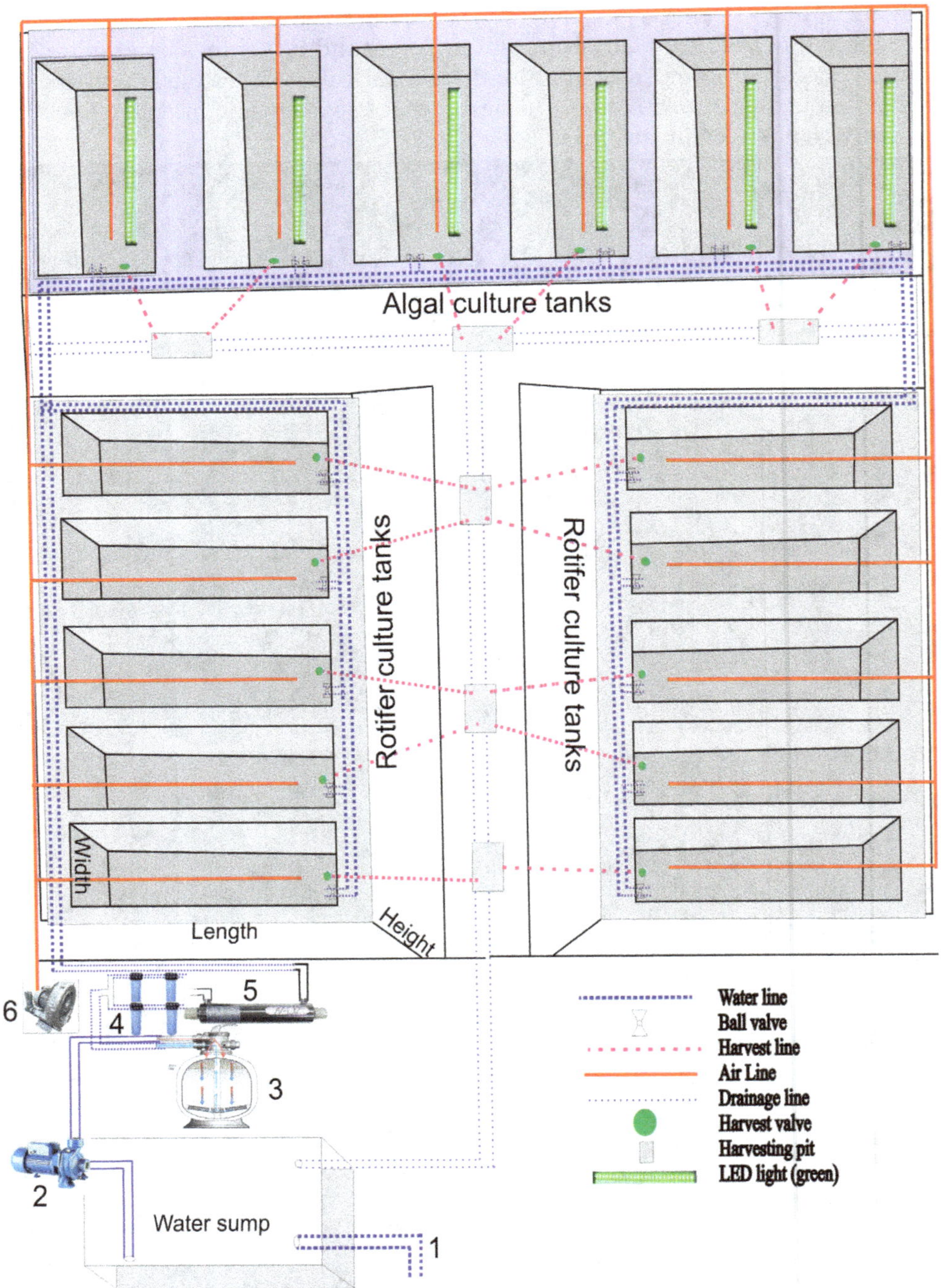

Plate 31: Layout of Live food production unit with RAS facility

1 – Main water inlet; 2 – Motor pump; 3 – Sand filter; 4 – Cartridge filters arranged in series; 5 – UV filter and 6 – Air blower

Plate 32: Cartridge and U.V. filtration setup (in-series)

Plate 33: Sand filter, Air blower and water pump attached at life food unit

Plate 34: Rotifer, *Brachionus calyciflorus,* culture tanks with under wateraeration facility

3.5 Formulation of probiotic incorporated encapsulated feed

3.5.1 Isolation, identification and characterization of *Lactobacillus sp.* from selected fish intestine as a probiotic

3.5.1.1 Collection of fish intestine sample

A total of 30 live and healthy adult fish samples of *D. dangila* (15) and *P. sophore* (15), with an average weight of 5.2±0.35 g and 4.25±0.24 g respectively, were sacrificed by overdosing of anesthetic (MS222) and washed with sterile water. The fish bodies were disinfected by 70% alcohol and dissected under antiseptic conditions. Intestines taken out were washed thoroughly with normal saline (NaCl 0.85 % w/v) to remove unwanted matters. Approximately 1 g of intestinal content was homogenized in 9 mL saline solution for centrifugation (Rengpipat et al., 2008). After centrifugation, the supernatant was serially diluted in the test tubes using sterile distilled water at 10^{-2}, 10^{-3}, 10^{-4}, 10^{-5}, 10^{-6} and 10^{-7} cfu mL^{-1} dilutions. The dilutions were then pour plated on de Man Rogosa and Sharp (MRS) agar (Merck) plates and incubated anaerobically (Oxoid anaerobic gas pack jar) for 48 hrs at 37°C temperature (Plate 35).

3.5.1.2 Isolation of lactic acid bacteria (LAB) from fish gut

Individual colonies with typical characteristics like pure white, off white, Yellow, yellow white, small (2–3 mm diameter) with entire margins were picked from each plate, transferred to subculture in three times and quadrant streak for pure culture in single colony. Selective colonies were characterized and identified following Bergey's Manual of Systematic Bacteriology (Whitman et al., 2009) for colony and cell morphology, gram staining, biochemical and physiological tests (Ghosh et al., 2002) (Plate 36 and 37).

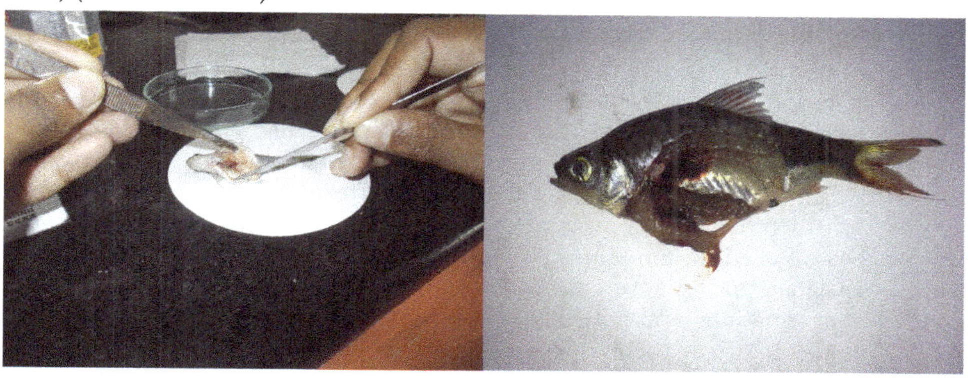

Plate 35: Collection of intestine from *Danio dangila* and *Puntius sophore*

3.5.1.3 Biochemical tests for LAB identification

a) **Antagonistic effect of the isolate**

Antibacterial activity of the strains was tested by disc and well diffusion techniques with cell free culture broth of the colonies selected (Cappuccino and Sherman, 2002). *Aeromonas hydrophila* (2µg/disc) was cultured in Tryptic Soy Broth (TSB) by streaking on TSA plates and incubated at 30°C for 24 hrs. During same time, the isolates were cultured in MRS broth for 24 hrs at 30°C and the bacterial cells were precipitated at 8000 rpm for 5 min (4°C). The supernatant was used for further investigation. The sterile discs were immersed in the supernatant and placed on TSA plates containing *A. hydrophila* after air drying. The plates were incubated at 30°C for 48 hrs to observe inhibition zone (Aly et al., 2008; Lauzon et al., 2008 and Rengpipat et al., 2008). One of the isolates with the greatest inhibition zone was selected for more potential probiotic evaluation.

b) pH tolerance

Acidification of the selected bacterium was investigated at different pH. MRS broths with different pH including 3, 4, 5, 6, 7, 8 and 9 were prepared using HCl 1% and NaOH 1 N and divided in universal bottles (Samelis et al., 1994). The broths media was sterilized at 121°C for 15 min and inoculated with overnight culture of the selected strain in MRS broth followed by incubation at 30°C. Optical density (OD) as growth rate of bacteria was measured by spectrophotometer at 600 nm after 2 h incubation. (Balcazar et al., 2008; Kim and Austin, 2008 and Allamesh et al., 2012).

c) Bile salt tolerance

Bile salt tolerance was further tested in MRS broth which included 0.0, 0.15 and 0.3% (w/v) Oxgall bile salt. Bottles of MRS broth containing different concentrations of bile salt were inoculated by 30 µl of cultured strain and incubated at 30°C. Growth rate was assessed by measuring the optical density by spectrophotometer at 600 nm after 0, 2, 4 and 8 h incubation (Balcázar et al., 2008; Kim and Austin, 2008 and Allamesh et al., 2012).

d) Antibiotic sensitivity test of the isolate

Antibiotic sensitivity test was carried out for selected strain by using common antibiotics in aquaculture by disc diffusion technique which included Tetracycline: 5 µg disc^{-1}, Penicillin: 2 µg disc^{-1}, Erythromycin: 5 µg disc^{-1}, Chloramphenicol: 10 µg disc^{-1}, Ofloxacin: 5 µg disc^{-1}, Ciprofloxacin: 5 µg disc^{-1}, Ampicillin: 2 µg disc^{-1} and Ceftazidime: 5 µg disc^{-1}. Around 50 µL culture isolate was spread on MRS agar and bio-discs (antibiotic) were placed on the plates. The plates were incubated at 30°C for 24 hrs to detect and quantify the inhibition zone (Kim and Austin, 2008). The interpretations and zone sizes were illustrated based on table of Kirby-Bauer test (Bauer et al., 1966) (Plate 38).

e) Catalase biochemical test

A small amount of isolate was placed on the microscopic slide. Using a Pasteur pipette, 1 drop of 3% Hydrogen peroxide (H_2O_2) was placed onto the isolate on the slide. Immediately the petri dish was covered with a lid followed by observation of immediate bubble formation (O_2 + H_2O = bubbles). Positive reactions were evident by immediate froth (bubble) formation.

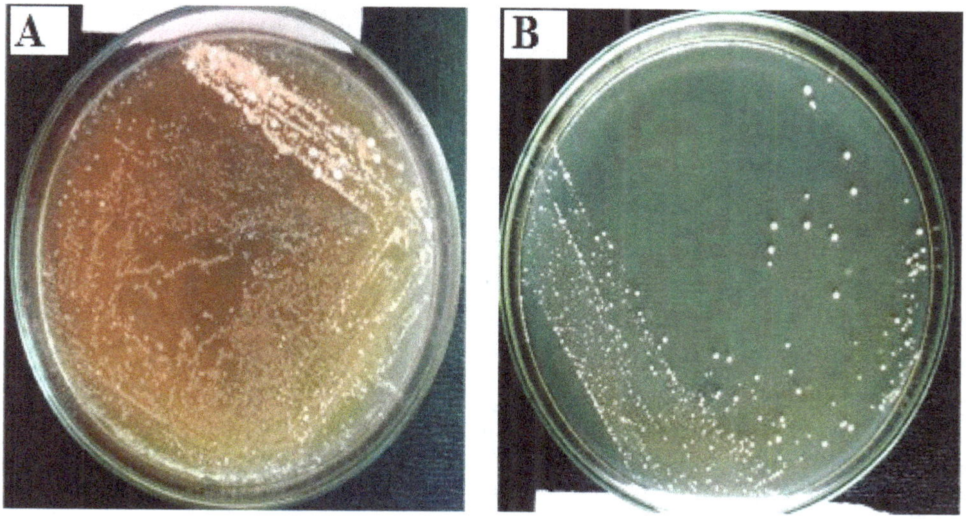

Plate 36: A - Isolated Lactobacillus sp. colonies from *D. dangila* B - Screened colonies on MRS from *D. dangila*

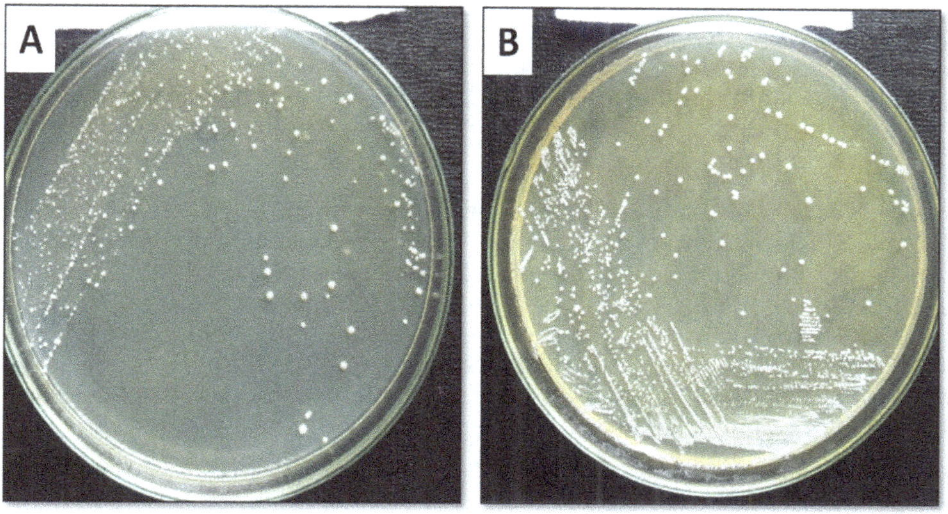

Plate 37: A - Isolated *Lactobacillus sp.* colonies from *P. sophore* B - Screened colonies on MRS from *P. sophore*

f) Indole biochemical test

Samples were inoculated in peptone water with amino acid tryptophan and incubated at 37°C for 24 hours. After incubation, few drops of Kovacs reagent (para-dimethyl amino benzaldehyde, isoamyl alcohol and conc.

HCl) were added. Results were concluded based on formation of a red or pink colour indicating the positive reaction.

g) **Oxidative fermentation test**

Oxidative fermentation (OF) test was used to determine if gram-negative bacteria metabolize carbohydrates oxidatively, by fermentation, or are nonsacchrolytic. Two tubes of Hugh and Leifson's OF basal medium were inoculated with the isolate using a straight wire by stabbing half way to the bottom of the tube. One tube of each pair was covered with 1 cm layer of liquid paraffin leaving the other tube open to the air. The tubes were incubated at 35°C for 48 hours and results were observed as fermentative, non saccharolytic or oxidative.

h) **Gram staining**

The glass slides were thoroughly cleaned, made grease free by alcohol (95%) and allowed to air dry. A drop of the culture was transferred on to the slide using inoculating loop and spread to form a thin film and air dried. The smear was fixed by passing the slide rapidly over the flame (film-side up). The smears were stained with crystal violet solution for 1 min and were rinsed with water. The Gram's iodine solution was applied and allowed to react (1 min) followed by rinsing with water. Alcohol, decolourizing solution, was added drop wise onto the smear until alcohol drippings from the slide were no longer coloured and were rinsed with water. The slides were then flooded with a counterstain, safranin, and allowed to react for 20-30 second followed by rinsing and air-dried. The stained smears were examined under the oil- immersion objective of the Phase Contrast microscope.

3.5.1.4 Mass culturing of Lactic Acid Bacteria (LAB)

The medium used in mass production of probiotic bacteria (LAB) was optimized as stated by Elmarzugi et al., (2010) and which was composed of yeast extract 6.0 g L^{-1}, peptone 6.0 g L^{-1}, potassium dihydrogen phosphate 1.5 g L^{-1}, ammonium citrate 1.0 g L^{-1}, tween 80 1.0 g L^{-1}, sodium acetate 1.0 g L^{-1}, citric acid 0.5 g L^{-1}, Magnesium sulphate heptahydrate 0.4 g L^{-1} and 0.05 g L^{-1}. The pH was adjusted to 6.8 before sterilization. Glucose, at 30 g L^{-1} concentration, was sterilized separately and added to medium before inoculation (Plate 39 and 40).

By following shake flask method, cultivation was carried out using 500 ml Erlenmeyer flasks containing 100 mL liquid medium. Inoculum was inoculated to the production medium followed by incubation at 37°C on rotary shaker at 200 rpm for 24 hrs. pH was measured using pH electrode and controlled at constant value by addition of 4M Sodium hydroxide (NaOH) or 2M hydrochloric acid (HCL) solutions. Dissolved oxygen concentration was maintained within acceptable limits. Antifoam reagent (Silicon antifoam, Merck) was added to overturn foaming during the experiment.

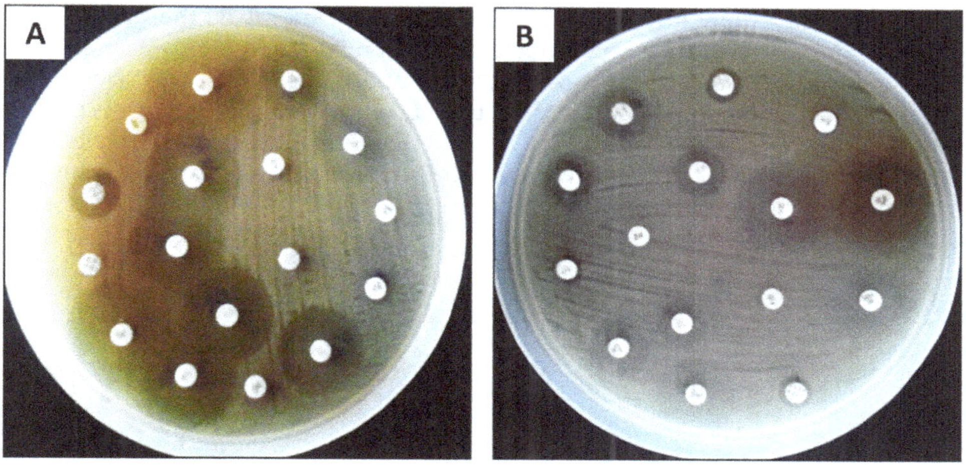

Plate 38: Antibiotic susceptibility test of isolates: **A.** *D. dangila* and **B.** *P. sophore*

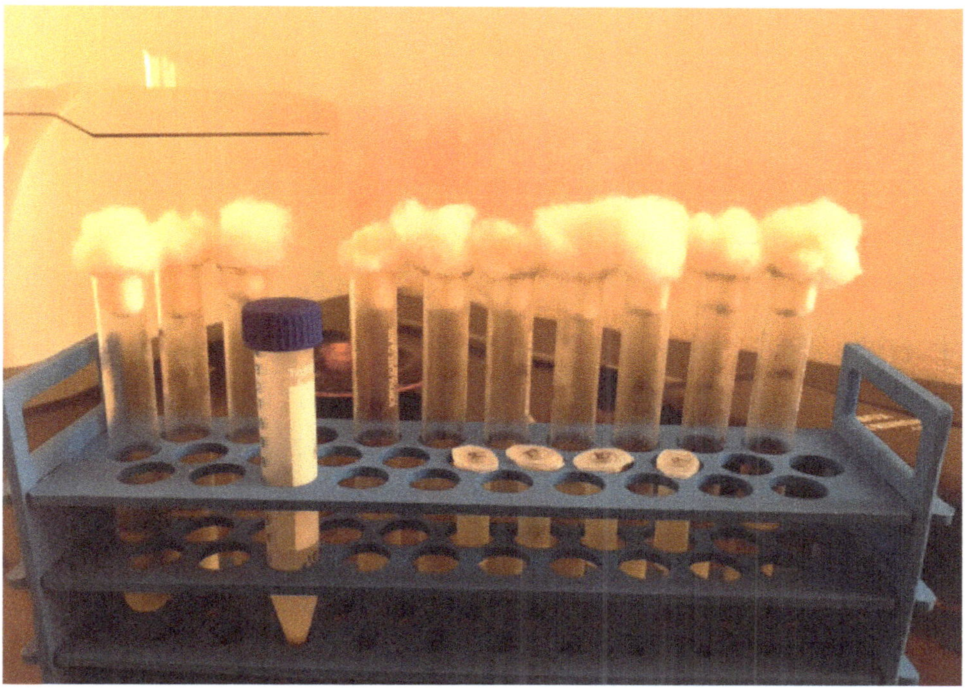

Plate 39: Stock preparation of *Lactobacillus spp.*

3.5.2 Feed preparation

All the feed ingredients of the experimental diet were dried, powdered and sieved (100μ) to get fine powder, mixed by using domestic mixer and made into dough like consistency by adding water (40%). Details of ingredients used are listed in Table 11. The dough was steamed in an autoclave at 100°C for five minutes and then cooled to room temperature. The cooled dough was hand mixed thoroughly after dispensing the lipid sources and vitamin-mineral mix. The dough was pelletized through a 2 mm diameter and dried overnight at 55°C using a hot air oven. After drying the pellets were stored at - 5° C until use. The dried pellets were broken into small pieces of about 500μm and stored separately in air tight containers to avoid microbial infection. Proximate composition of feed was analyzed following standard methods (AOAC, 2012) (Plate 41).

Table 11: Ingredients used for feed preparation and proximate composition

Sl. No.	Ingredient	Quantity used (%)
1.	Fish meal	40.00
2.	Shrimp meal	10.00
3.	Wheat flour	10.00
4.	Soyabean flour	15.00
5.	Corn powder	10.00
6.	Vitamin E (Fish oil)	05.00
7.	Vit-Min premix	01.00
8.	Sodium based montmorelonite	0.50
	Proximate composition	
1.	Crude protein	50%
2.	Moisture	04%
3.	Ash	8.4%
4.	Crude Lipid	14%
5.	Crude Fibre	1.5%
6.	Gross energy (kcal g^{-1})	7.85

3.5.3 Encapsulation of probiotic bacteria in feed

Encapsulation of *Lactobacillus sp.* was done as per the method reported by Xing et al., (2014). *Lactobacillus sp.* was activated on MRS broth at 37°C for 24 hours. The cells were harvested by centrifugation at 5°C with 3000 rpm for 10 min. The obtained cells were washed twice and adjusted to the concentration of 10^1 to 10^7 cfu mL^{-1}. The cell suspension was added to the feed at 10^1 cfu g^{-1}, 10^3 cfu g^{-1}, 10^5 cfu g^{-1} and 10^7 cfu g^{-1}. After absorption, sodium alginate with 2% concentration and soybean oil (10 mL) were added drop wise during stirring. In order to harden the coating material, 0.1M calcium chloride (100 mL) was added and refrigerated immediately at -40°C for 2 hours and air dried. After air drying, the feed samples with different concentrations of *Lactobacillus sp.* (cfu g^{-1}) were stored at room temperature and -20°C for LAB viability analysis.

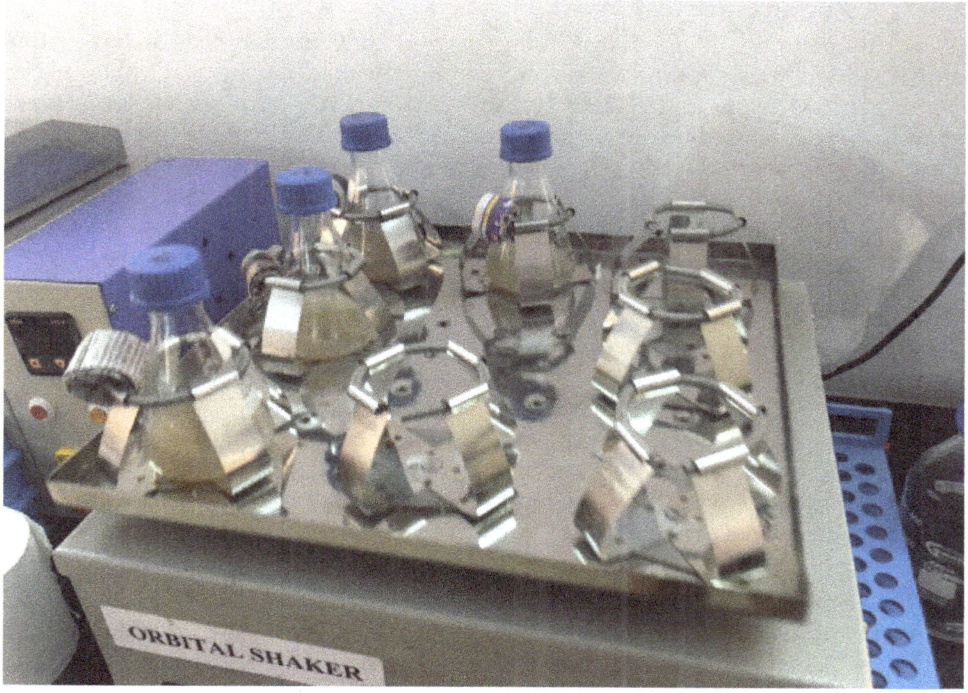

Plate 40: Mass production of *Lactobacillus spp.*

Plate 41: Probiotic incorporated encapsulated feed with different concentrations

3.5.4 LAB occurrence test

Approximately 1 g of feed was homogenized in 9 mL saline solution. After centrifugation, the supernatants were serially diluted in the test tubes using sterile distilled water at 10^{-2}, 10^{-3}, 10^{-4}, 10^{-5}, 10^{-6} and 10^{-7} dilutions. The dilutions were then pour plated on de Man Rogosa and Sharp (MRS) agar (Merck) plates and incubated anaerobically (Oxoid anaerobic gas pack jar) for 48 hours at 37°C temperature (Plate 42). The number of colony forming was calculated under colony counter and LAB viability was analyzed using following mathematical expression.

3.5.5 Effect of probiotic incorporated encapsulated feed on larval growth

Probiotic bacteria (*Lactobacillus sp.*) incorporated encapsulated feed was fed to selected ornamental fish larvae (*D. dangila* and *P. sophore*). The experiment was performed following Completely Randomized Design (CRD). Probiotic incorporated encapsulated feed with different concentration (10 cfu g^{-1}, 10^3 cfu g^{-1}, 10^5 cfu g^{-1} and 10^7 cfu g^{-1}) was applied to fish larvae in triplicate at 8-10% of body weight thrice a day (Table 12).

The bio-growth parameters were analyzed by estimating final weight gain (gm), specific growth rate (SGR) (%), average daily gain (ADG) (g), feed conversion ratio (FCR) and survival rate (%) using following mathematical expressions.

Final weight gain = *Final weight* (L_1) – *Initial weight* (L_0)

Specific Growth Rate (SGR) (%) = *In Final weight* (L_1) – *In Initial weight* (L_0)

(Hewette et al., 1991)

$$\textbf{Average Daily Gain (ADG)} = \frac{\text{Mean final weight} - \text{Mean initial weight}}{\text{Time } (t_1 - t_0)}$$

(Hawkins et al., 1985)

$$\textbf{Feed Conversion Ratio (FCR)} = \frac{\text{Feed consumed by fish (g)}}{\text{Weight gain (Final weight} - \text{Initial weight) (g)}}$$

(Hewette and Johnson, 1992)

$$\textbf{Survival rate (\%)} = \frac{\text{Total number of fish counted}}{\text{Total number of fish stocked}}$$

(Cheikyula and Ofojekwo, 2003)

Table 12: Effect of probiotic incorporated encapsulated feed on selected ornamental fish larvae

Treatment	T_c	T_1	T_2	T_3	T_4
Probiotic encapsulated feed (cfu g^{-1})	00	1×10^1	1×10^3	1×10^5	1×10^7

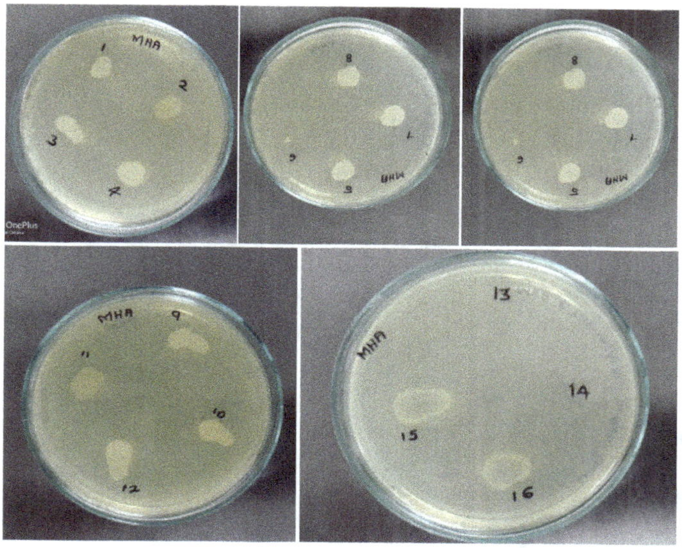

Plate 42: LAB occurrence test

3.5.6 Histological analysis of selected fish species

For histological characterization, 15 fish larvae from each experiment were selected and fixed in buffered 10% formalin. After dehydration and clearing, fish larvae were embedded in paraffin wax followed by sectioning on a microtome at a thickness of 5 μm and stained with Haematoxylin and Eosin (Plhalova et al., 2012). From each larva 10 slides were produced for microscopic analysis. Histopathological changes such as whole larvae, intestinal villi length (μm), number of villi, muscular thickness and other possible morphological changes were evaluated and compared with the control group.

3.6 Larval rearing of selected ornamental fishes

3.6.1 Effect of rearing system and stocking density on larval rearing

Larval rearing experiment was conducted following 2×4 factorial design. Selected ornamental fishes (*D. dangila* and *P. sophore*) were stocked in two rearing systems (Plate 43 and 44) *viz.* nursery raceways and FRP tanks with four different stocking densities *viz.* 10 nos. L^{-1}, 30 nos. L^{-1}, 50 nos. L^{-1} and 70 nos. L^{-1} in triplicate (Table 13). Nursery rearing raceways (10000 L) (6×2×1.15 m) were composed of separate inlet and outlet with Recirculatory Aquaculture System (RAS) and adjustable air lift systems for aeration and continuous water flow (Plate 45 and 46). Whereas, FRP tanks (1000L) were utilized following traditional culture technique. Spawn (4 DPH) were fed initially with Rotifers, *Brachionus calyciflorus*, followed by supplementary feed at 8-10% of body weight twice a day. The physico-chemical parameters of water were kept at optimal level during the experiment. The results were obtained by analyzing final weight gain, average daily gain (ADG), specific growth rate (SGR) and survival rate (%) as expressed by following mathematical expressions as explained in section 3.6.4 (Plate 47 and 48).

Plate 43: Spawns of one day old

Plate 44: Spawn of three day old

Plate 45: RAS based nursery raceways

Plate 46: Filtration facility (sand and U.V. filters) at nursery raceway

Plate 47: Rearing of *Danio dangila* in nursery raceways

Plate 48: Rearing of *Puntius sophore* in nursery raceways

Table 13: Effect of rearing system and stocking density on larval rearing of selected ornamental fish species

Rearing systems	Treatment	T_1	T_2	T_3	T_4
1. Nursery Raceways 2. FRP tanks	Stocking density (nos. L^{-1})	10	30	50	70

3.7 Statistical analysis

Initially experimental data was tested for normality of distribution using Shapiro-Wilk's W-statistic (PAST, version 16.0). One-way analysis of variance (ANOVA) was used to compare significant differences between treatments. After getting significant differences of triplicate mean with 5% level of significance ($p<0.05$), Duncan's multiple range test (Duncan, 1995) was used to rank the groups on SPSS software, version 20.0 (SPSS Inc., Michigan Avenue, Chicago, Illinois, U.S.A.). The results are presented as mean ± SEM. In the experiment of stocking density optimization, data of FRP tanks and Raceways were compared and analyzed following two way analysis of variance (ANOVA) in SPSS software, version 20.0 (SPSS Inc., Michigan Avenue, Chicago, Illinois, U.S.A.).

CHAPTER 4

Results

4.1 Induced breeding of Moustached danio, *Danio dangila* (Hamilton, 1822)

The number of breeding trials, size of brooders, latency period, number of eggs spawned, fertilization rate, hatching rate and consequent remarks using different hormones are described in Table 14. Matured females were larger when compared with males in the selected species while males were brighter, attractive and aggressive compared with females.

4.1.1 Breeding behavior

Brooders showed mating behavior after inducement of different hormones in most of the treatments, except in the control and low dosage sets. Mating was preceded by elaborated courtship behavior. First male started chasing the female followed by releasing milt over the eggs released by female in water. Brooders of *D. dangila* showed breeding behavior after 2.0-2.5 hours of hormonal injections. In case of T_0 and T_1, no breeding behavior was seen in all groups while in case of high dosage administration (T_4), female settled down at bottom with slow movement and stressed.

4.1.2 Latency period

Latency period varied significantly between the hormones and dosages injected to the brooders of the selected ornamental fish (Table 14). Almost all fishes have spawned within 6-8 hours of hormonal injection, except with Ovatide in T_3 as spawning took place as early as after 6 hours of injection. Latency period was delayed and seen near to 8:00 hours (8:82 hours) of WOVA-FH and HCG injection (T_2). The effect of dosages on mean latency period (hours) has been illustrated in Fig. 1.

4.1.3 Number of eggs spawned

The number of eggs released varied significantly between the hormones and doses injected to the females of *D. dangila* (Table 14). Significantly highest ($p<0.05$) number of spawned eggs was observed in T_3 (1002±52.6) followed by T_2 (675.50±52.61) when fishes were injected with WOVA-FH. Subsequently, the higher number of spawned eggs was observed when fishes were injected with Ovatide in T_2 (910±39.2) followed by T_3 (753.0±1.09). Nonetheless, when fishes were injected with HCG, lower number of eggs was observed in both spawned sets T_2 (361.0±1.80) and T_3 (354.0±2.21). The effect of hormonal dosages on mean number of eggs spawned by *D. dangila* has been illustrated in Fig. 2.

4.1.4 Fertilization rate

The fertilized eggs of *D. dangila* were spherical and translucent with pale brownish colour measuring around 0.8 mm to 1.0 mm diameter. Unfertilized eggs were pale, opaque and white in colour. Fertilization rate in *D. dangila* was significantly higher (*p<0.05*) at T_3 (84.0±2.72%) followed by T_2 (72.0±3.39%) when injected with WOVA-FH. Subsequently, when the fishes were injected with Ovatide, higher fertilization rate was observed in T_2 (68.0±2.85%) trailed by T_3 (60.5±2.70%). However, the lower fertilization rate was observed when the fishes were injected with HCG showing at T_2 (67.0±5.90%) and T_3 (54.0±9.60%). Highest fertilization rate was observed at 0.5 mL kg^{-1} for female and 0.3 mL kg^{-1} for male (84.0±2.72%) using WOVA-FH than other hormones and doses. The effect of hormonal dosages on mean fertilization rate of *D. dangila* has been illustrated in Fig. 3.

4.1.5 Hatching rate

Twitching movement of the embryos was observed within 60:00 hours of spawning and the young ones hatched out within 68–72 hours (Table 13) at 25.5±1.5°C temperature. Hatching rate was varied significantly (*P<0.05*) between the hormones and doses injected to the females of *D. dangila* (Table 14). Significantly higher hatching rate in *D. dangila* was observed at T_3 (86.5±2.9%) followed by T_2 (75.0±3.0%) when induced with WOVA-FH. Subsequently, when fishes injected with Ovatide, T_2 showed higher hatching rate (75.5±3.2%) followed by T_3 (59.5±2.3%). But when fishes were induced with HCG, consequently lower hatching rates were observed at T_2 (67.0±5.90%) followed by T_3 (54.0±9.60%). Higher hatching rate was observed in *D. dangila* (86.5±2.9%) at 0.5 mL kg^{-1} for female and 0.3 mL kg^{-1} for male using WOVA-FH when compared with other hormones and doses. The effect of hormonal dosages on mean hatching rate of *D. dangila* has been illustrated in Fig. 4. Chronological development of embryo of the species has been described in Table 15 and presented in Plate 17.

Table 14: Effect of different hormones and dosages on reproductive parameters of *Danio dangila* during induced breeding

Treatment	Male weight (g)	Female weight (g)	Latency Period (Hours)	Number of eggs spawned	Fertilization (%)	Hatching (%)	Remarks
WOVA-FH							
T_0	5.25 ± 0.95	6.57 ± 0.20	0^a	0^a	0^a	0^a	No spawning
T_1	4.75 ± 0.30	6.50 ± 0.14	0^a	0^a	0^a	0^a	No spawning
T_2	5.02 ± 0.34	6.30 ± 0.19	8.82 ± 0.56^c	675.5 ± 52.61^b	72.0 ± 3.39^{bc}	75.0 ± 3.0^b	Partial spawning
T_3	4.8 ± 0.23	6.60 ± 0.23	7.87 ± 0.12^b	1002.50 ± 52.6^c	84.0 ± 2.72^c	86.5 ± 2.9^{bc}	Complete spawning
T_4	4.9 ± 0.31	6.6 ± 0.23	0^a	0^a	0^a	0^a	No spawning

p- value	-	-	0.024	0.025	0.037	0.026	
HCG							
T_0	5.15 ± 0.12	6.42 ± 0.25	0^{a}	0^{a}	0^{a}	0^{a}	No spawning
T_1	4.85 ± 0.26	6.67 ± 0.18	0^{a}	0^{a}	0^{a}	0^{a}	No spawning
T_2	5.55 ± .0.52	6.55 ± 0.20	8.82 ± 0.56^{c}	361 ± 1.80^{b}	67.0 ± 5.90^{b}	68.5 ± 4.99^{b}	Partial spawning
T_3	5.07 ± 0.40	6.55 ± 0.21	7.37 ± 0.84^{bc}	354 ± 2.21^{b}	54.0 ± 9.60^{b}	59.2 ± 9.90^{b}	Partial spawning
T_4	4.90 ± 0.63	6.46 ± 0.21	0^{a}	0^{a}	0^{a}	0^{a}	No spawning
p- value	-	-	0.025	0.046	0.026	0.024	
Ovatide							
T_0	5.45 ± 0.12	6.78 ± 0.36	0^{a}	0^{a}	0^{a}	0^{a}	No spawning
T_1	5.05 ± 0.15	6.33 ± 0.19	0^{a}	0^{a}	0^{a}	0^{a}	No spawning
T_2	5.45 ± 0.33	6.55 ± 0.14	8.0 ± 0.40^{bc}	910 ± 39.20^{b}	68.0 ± 2.85^{b}	75.5 ± 3.2^{c}	Partial spawning
T_3	4.67 ± 0.18	6.75 ± 0.13	6.2 ± 2.04^{b}	753 ± 1.09^{b}	60.5 ± 2.70^{b}	59.5 ± 2.3^{b}	Partial spawning
T_4	5.37 ± 0.38	6.37 ±.0.13	0^{a}	0^{a}	0^{a}	0^{a}	No spawning
p- value	-	-	0.024	0.025	0.037	0.026	

***Note:** Values with different superscripts in a column differs significantly at $p<0.05$ (n=3)*

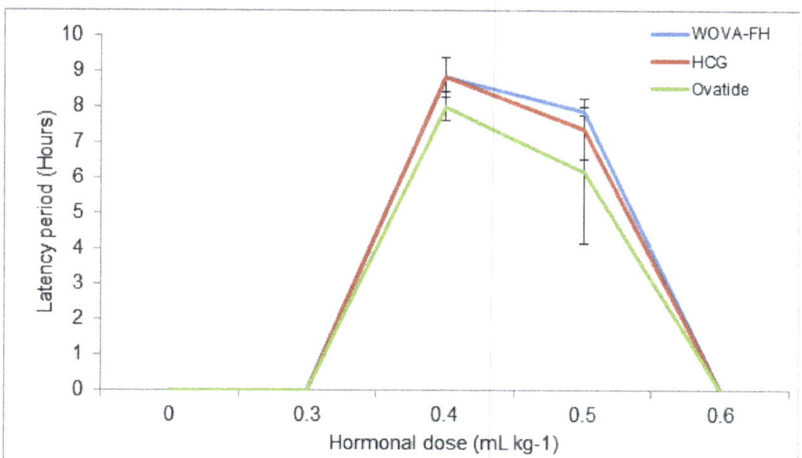

Fig. 1: Effect of different hormones and dosages on mean latency period (hours) in *D. dangila*

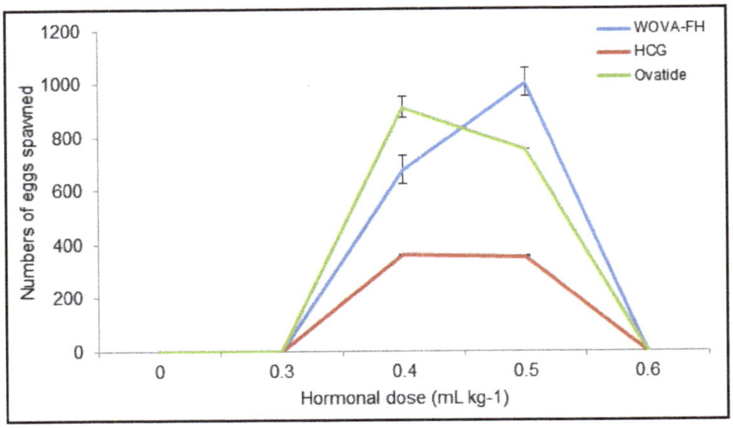

Fig. 2: Effect of different hormones and dosages on mean number of eggs spawned (eggs female^{-1}) in *D. dangila*

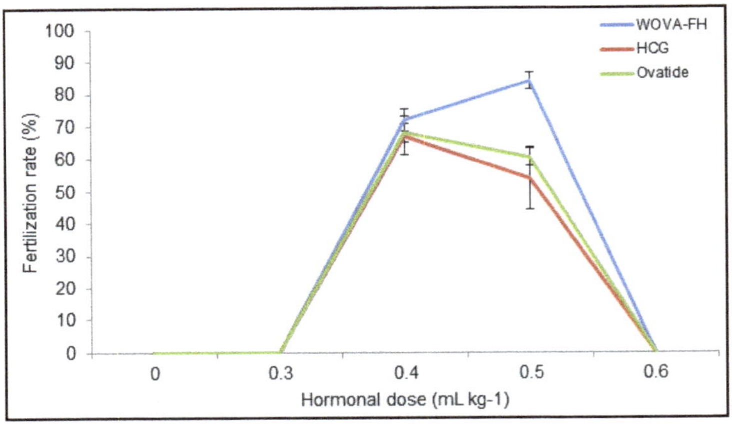

Fig. 3: Effect of different hormones and dosages on mean fertilization rate (%) in *D. dangila*

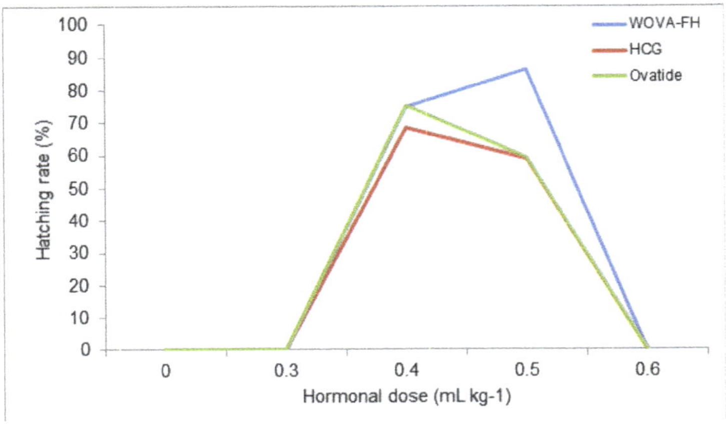

Fig. 4: Effect of different hormones and dosages on mean hatching rate (%) in *D. dangila*

Table 15: Description on embryonic development of *Danio dangila* recorded during captive breeding experiment (Plate 49)

Developmental Stage	Descriptional characteristics	Mean Development time (Hours:min.)
Zygote period (1 cell)	Yolk-free cytoplasm begins to stream towards fish pole segregating blastodisc. Blastodiscs were not observable.	00:20
Cleavage Period (2 cells)	Two blastomers were rounded.	00:45
Morula (64 cells)	Blastodisc consists of many blastomers.	01:45
Blastula (256 cells)	Blastodermal size reduced and number of marginal cells increased.	02:45
Oblong stage	A smoothy outlined ellipsoidal shape observed.	03:45
Sphere shape	Border between blastodisc and yolk flattened and spreaded gradually.	04:30
Dome shape	Blastoderm became dome like.	04:50
Gastrula	Gastrulation begins forming epiboly.	05:30
30% epiboly	Blastoderm became flattened down onto the yolk sac and germ ring epiboled 30% of yolk sac.	05:35
50% epiboly	Germ ring covered 50% of yolk sac and neural plate started forming.	06:00
100% epiboly	Yolk invasion completed.	06:30
Bud formation	Epiboly came close to blastoderm completely. Tail bud appeared.	11:00
Segmentation	First somite formed.	12:00
4 somites	4 somites visible with neurocoele cells in head.	14:00
8 somites	8 somites visible with optic vesicles and notochord.	16:00
12 somites	12 somites observable with prominent tail bud.	18:00
20 somites	20 somites visible. Heart beats were observable.	24:00
Pharyngula	Eyes clearly visible with bulgy yolk sac. Embryonic movement started.	38:00
	Pharyngula stage continued for another 26-28 hours.	-
Hatching	Fully transparent hatchlings was measured around 1.8-2.4 mm.	72:00
Newly hatched larvae	Head slightly bent on yolk. Responsive to stimulus.	

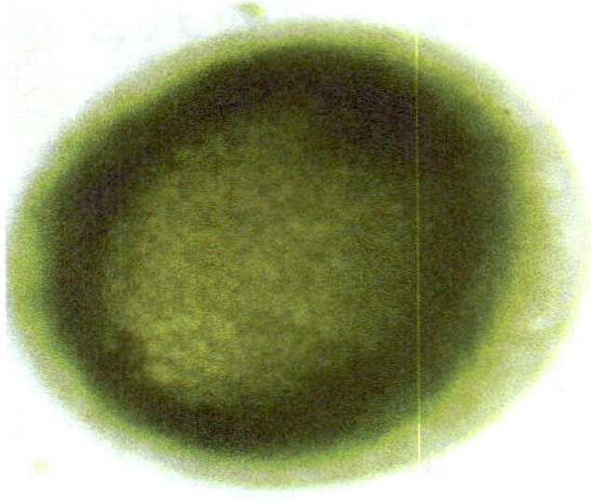

Fertilized egg

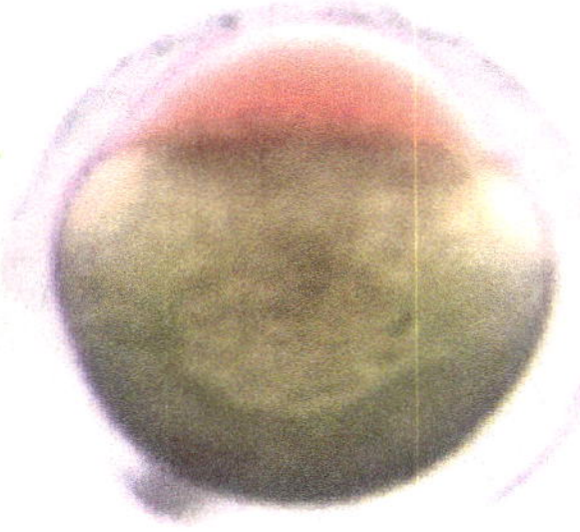

1-Cell stage

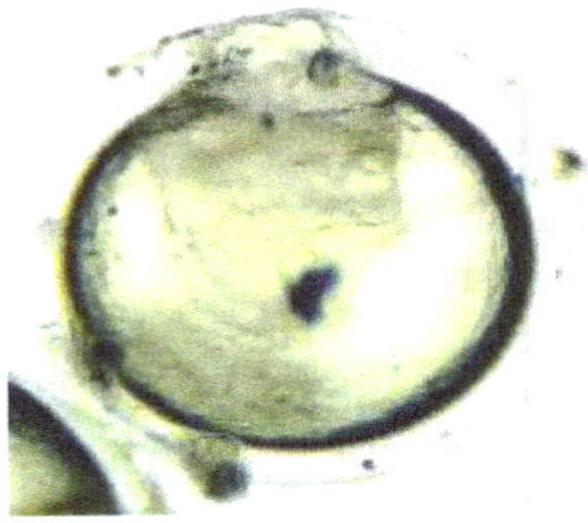

2-cell stage

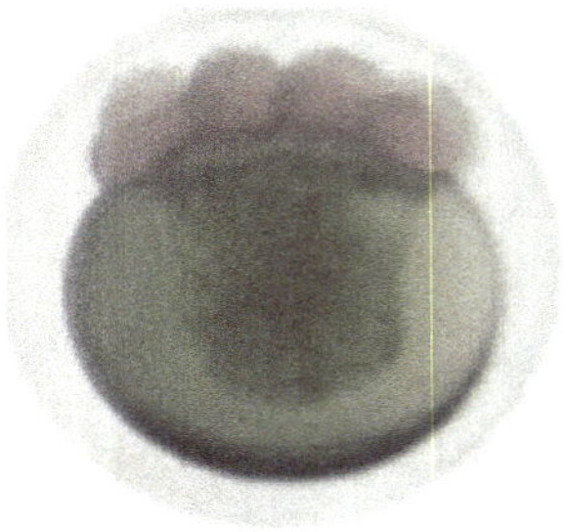

8-Cell stage

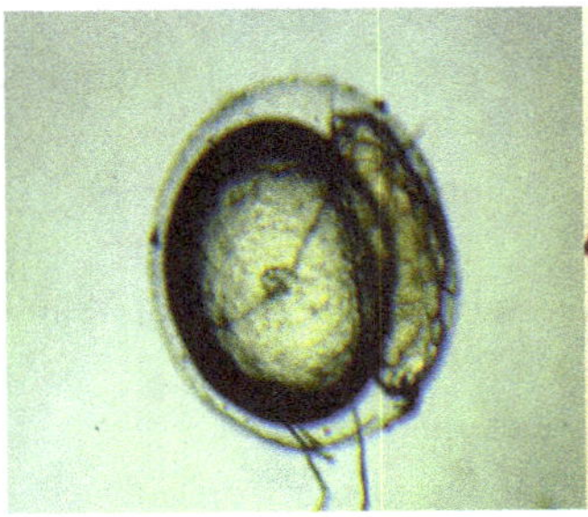

32-Cell stage

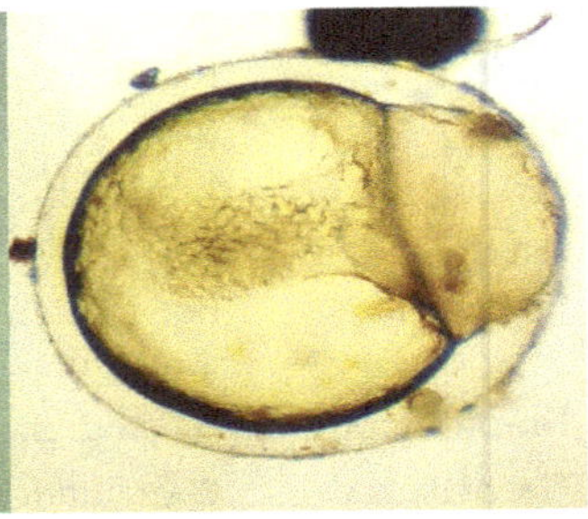

64-Cell stage

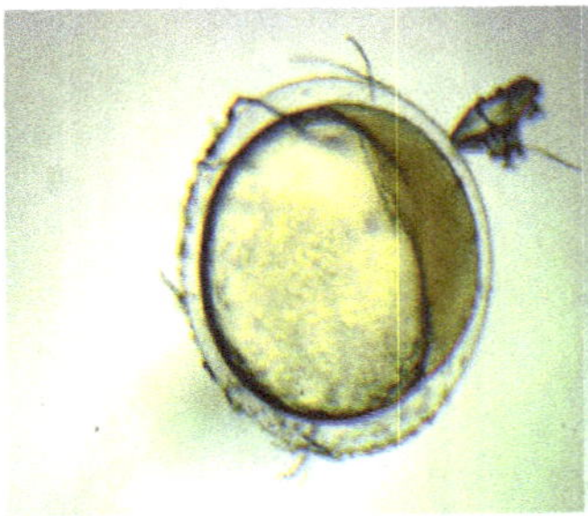

Morula stage

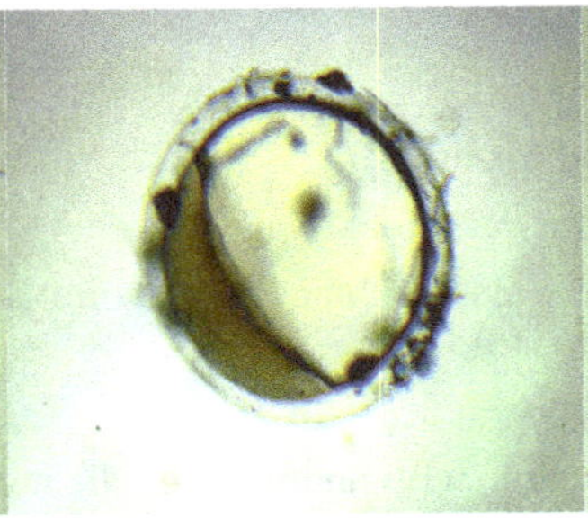

Blastula stage

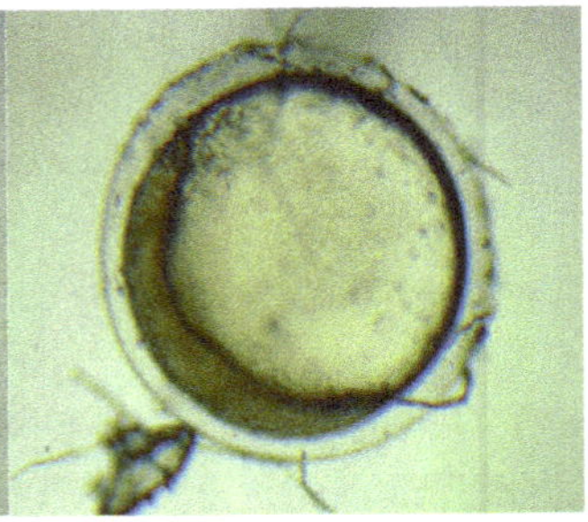

Gastrula stage

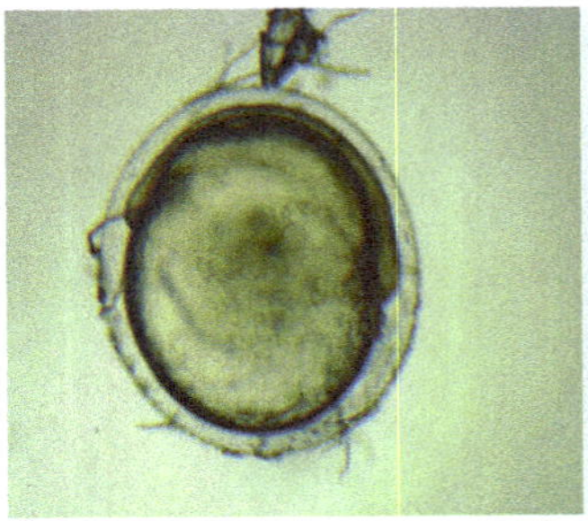

Embryonic stage

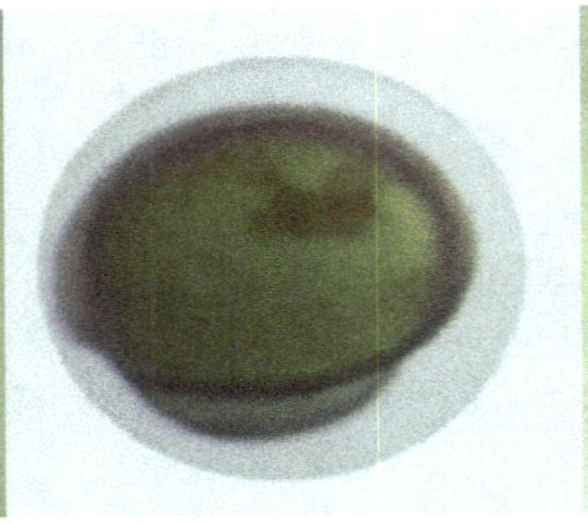

Epiboly stage

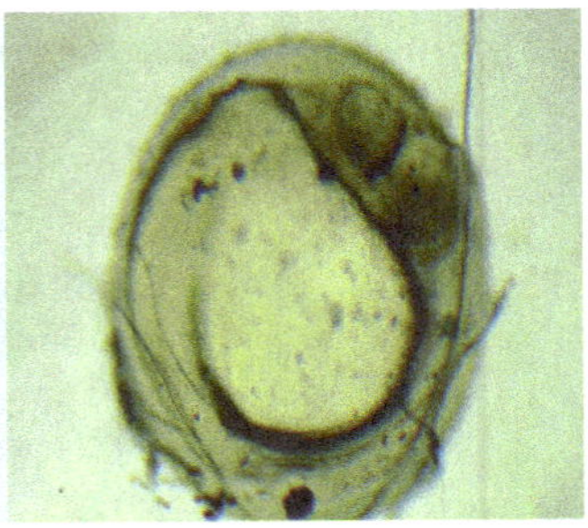

Somite formation stage

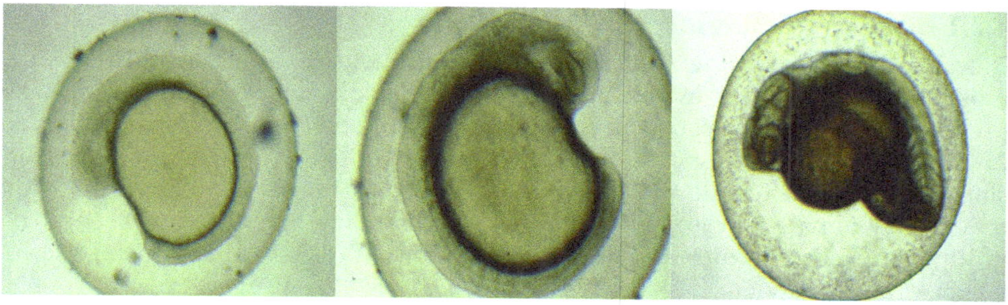

Somite formation (12-somite) | Heart formation | 15 Somite stage

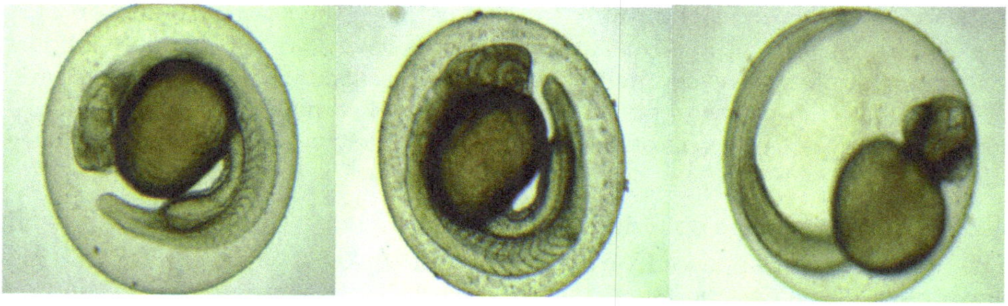

17 Somite stage | 23- Somite stage | Pharyngula stage

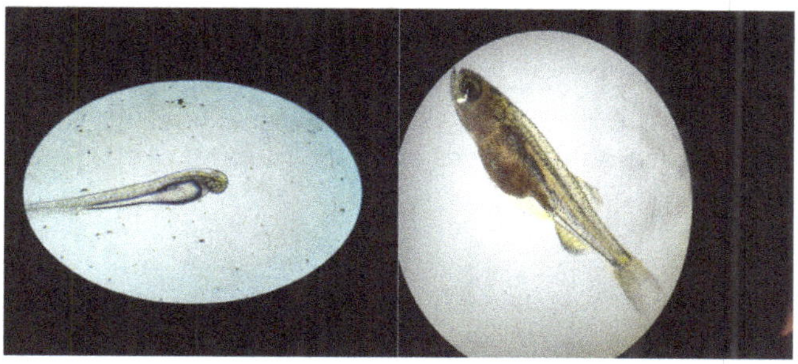

Larvae after hatching from chorion | Larvae after yolk sac absorption

Plate 49: Embryonic developmental stages recorded during captive breeding of *D. dangila*

4.2 Induced breeding of Sophore barb, *Puntius sophore* (Hamilton, 1822)

The number of breeding trials, size of brooders, latency period, number of eggs spawned, fertilization rate, hatching rate and consequent remarks using different hormones are described in Table 16. Matured females were larger when compared with males in the selected species while males were brighter with dark red horizontal patch parallel to vertebrae, attractive with green color on back and aggressive compared to females.

4.2.1 Breeding behavior

Brooders showed mating behavior after inducement of different hormones in most of the treatments, except in the control and low dosage sets. Mating was preceded by elaborated courtship behavior. Breeding behavior was observed after 2.0-2.5 hours of resolving dose to females and first dose to males. Initially, male started chasing the female and the courtship ended by releasing milt over the eggs released by female. In case of T_0, T_1 and T_4, no breeding behavior was seen in all groups induced with different hormones, except with WOVA-FH which showed late breeding behavior.

4.2.2 Latency period

Latency period varied significantly ($p<0.05$) between the hormones and dosages injected to *P. sophore* (Table 16). Almost all fishes have spawned within 6-8 hours of hormonal injection, except with Ovatide in T_3 as spawning took place as early as after 6 hours. But, in contrast, it was delayed and seen after 9 hours of WOVA-FH injection (T_4). The effect of dosages on mean latency period (hours) has been illustrated in Fig. 5.

4.2.3 Number of eggs spawned

The number of eggs released varied significantly ($p<0.05$) between the hormones and doses injected to the females of *P. sophore* (Table 16). Significantly highest ($P<0.05$) number of eggs was observed at T_2 (881.50±85.8) followed by T_3 (554.50±57.89) when fishes were induced with WOVA-FH. Subsequently, the higher number of spawned eggs was observed when fishes were injected with Ovatide in T_2 (710.0±39.2) followed by T_3 (553.0±10.9). Nonetheless, when fishes were injected with HCG, lower number of eggs was observed in T_2 (430.50±16.16) and T_3 (130.75±47.78). The effect of hormonal dosages on mean number of spawned eggs of *P. sophore* has been illustrated in Fig. 6.

4.2.4 Fertilization rate

The fertilized eggs of *P. sophore* look transparent with golden stinge measuring around 0.6 mm to 0.7 mm diameter. Unfertilized eggs were pale, opaque and white in colour. Fertilization rate in *P. sophore* was significantly higher ($P<0.05$) at T_2 (78.50±2.72%) followed by T_3 (66.0±3.39%) when injected with WOVA-FH. Subsequently, when the fishes were injected with Ovatide, highest fertilization rate was observed in T_2 (76.0±5.12%) trailed by T_3 (60.5±2.70%). However, the lower fertilization rate was observed when the fishes were injected with HCG at T_3 (66.0±5.60%) and T_2 (59.0±9.90%). Highest fertilization rate was observed at 0.4 mL kg^{-1} for female and 0.2 mL kg^{-1} for male (78.50±2.72%) using WOVA-FH than other hormones and doses. The effect of hormonal dosages on mean fertilization rate of *P. sophore* has been illustrated in Fig. 7.

4.2.5 Hatching rate

Twitching movement of the embryos was observed within 35:45 hours of spawning and the young ones hatched out within 38–40 hours (Table 17) at 25.5±1.5°C temperature. Hatching rate was varied significantly ($p<0.05$) between the hormones and doses injected to the females *P. sophore* (Table 16). Significantly higher hatching rate in *P. sophore* was observed at T_2 (83.75±2.39%) followed by T_3 (75.75±2.10%) when induced with WOVA-FH. Subsequently, when fishes injected with Ovatide, T_2 showed higher hatching rate (79.25±.49%) followed by T_3 (59.50±2.0%). But when fishes were induced with HCG, consequently lower hatching rates were observed at T_2 (60.5±2.10%) and T_3 (68.2±4.54%). Higher hatching rate was observed in *P. sophore* (83.75±2.39%) 0.4 mL kg^{-1} for female and 0.2 mL kg^{-1} for male using WOVA-FH when compared with other hormones and doses. The effect of hormonal dosages on mean hatching rate of *P. sophore* has been illustrated in Fig. 8. Chronological development of embryo of *P. sophore* has been described in Table 17 and presented in Plate 49.

Table 16: Effect of different hormones and doses on reproductive parameters of Puntius sophore during induced breeding

Treatment	Male weight (g)	Female weight (g)	Latency Period (Hours)	Number of eggs spawned	Fertilization (%)	Hatching (%)	Remarks
WOVA-FH							
T_0	4.15 ± 0.09	5.47 ± 0.20	0^a	0^a	0^a	0^a	No spawning
T_1	3.65 ± 0.30	5.40 ± 0.14	0^a	0^a	0^a	0^a	No spawning
T_2	3.92 ± 0.34	5.20 ± 0.19	7.87 ± 0.12^b	881.50 ± 85.8^d	78.50 ± 2.72^d	83.75 ± 2.39^d	Complete spawning
T_3	3.75 ± 0.23	5.50 ± 0.23	8.82 ± 0.56^c	554.50 ± 57.89^c	66.00 ± 3.39^c	75.75 ± 2.10^c	Partial spawning
T_4	3.80 ± 0.31	5.53 ± 0.23	9.12 ± 0.31^c	130.25 ± 1.30^b	16.75 ± 1.32^b	8.75 ± 1.23^b	Partial spawning
p- value	-	-	0.020	0.046	0.038	0.035	
HCG							
T_0	5.15 ± 0.12	5.22 ± 0.25	0^a	0^a	0^a	0^a	No spawning
T_1	4.85 ± 0.26	5.47 ± 0.18	0^a	0^a	0^a	0^a	No spawning
T_2	5.07 ± 0.20	5.32 ± .0.20	8.87 ± 0.96^b	430.50 ± 16.16^c	59.00 ± 9.90^b	60.50 ± 2.10^b	Partial spawning
T_3	5.55 ± 0.26	5.35 ± 0.21	8.07 ± 0.34^b	130.75 ± 47.78^b	66.00 ± 5.60^b	68.20 ± 4.54^b	Partial spawning
T_4	4.90 ± 0.31	5.26 ± 0.21	0^a	0^a	0^a	0^a	No spawning
p- value	-	-	0.044	0.030	0.038	0.031	
Ovatide							
T_0	4.25 ± 0.25	5.68 ± 0.36	0^a	0^a	0^a	0^a	No spawning

Treatment	Male weight (g)	Female weight (g)	Latency Period (Hours)	Number of eggs spawned	Fertilization (%)	Hatching (%)	Remarks
T_1	3.85 ± 0.15	5.23 ± 0.19	0^a	0^a	0^a	0^a	No spawning
T_2	4.25 ± 0.33	5.45 ± 0.14	7.87 ± 0.12^c	710.00 ± 39.20^c	76.00 ± 5.12^b	79.25 ± 1.49^b	Complete spawning
T_3	3.47 ± 0.18	5.55 ± 0.13	6.12 ± 0.20^b	553.00 ± 10.90^b	60.50 ± 2.7^b	59.50 ± 2.0^b	Partial spawning
T_4	4.17 ± 0.38	5.27 ± .0.13	0^a	0^a	0^a	0^a	No spawning
p- value	-	-	0.036	0.025	0.032	0.042	

Note: *Values with different superscripts in column differs significantly at p<0.05 (n=3)*

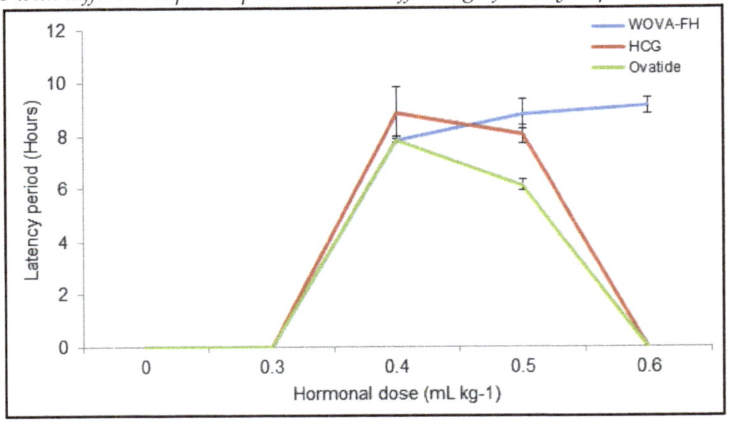

Fig. 5: Effect of different hormones and dosages on mean latency period (hours) in *P. sophore*

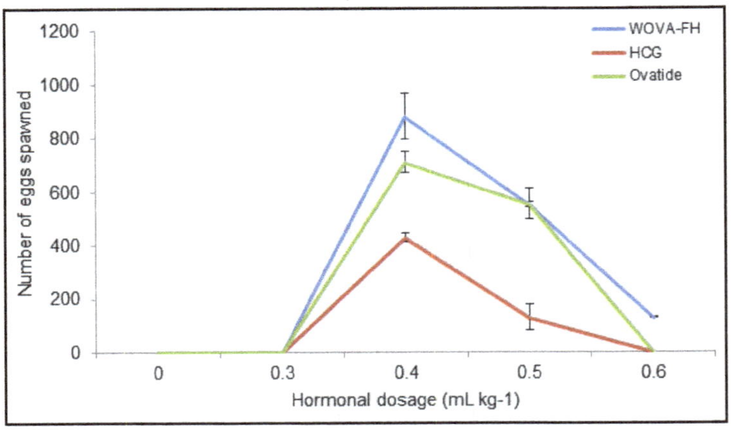

Fig. 6: Effect of different hormones and dosages on mean number of eggs spawned (eggs female^{-1}) in *P. sophore*

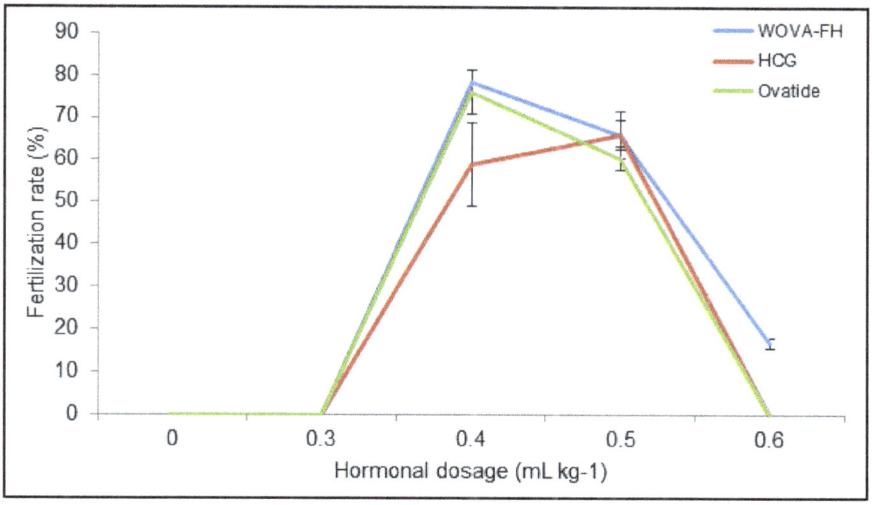

Fig. 7: Effect of different hormones and dosages on mean fertilization rate (%) in *P. sophore*

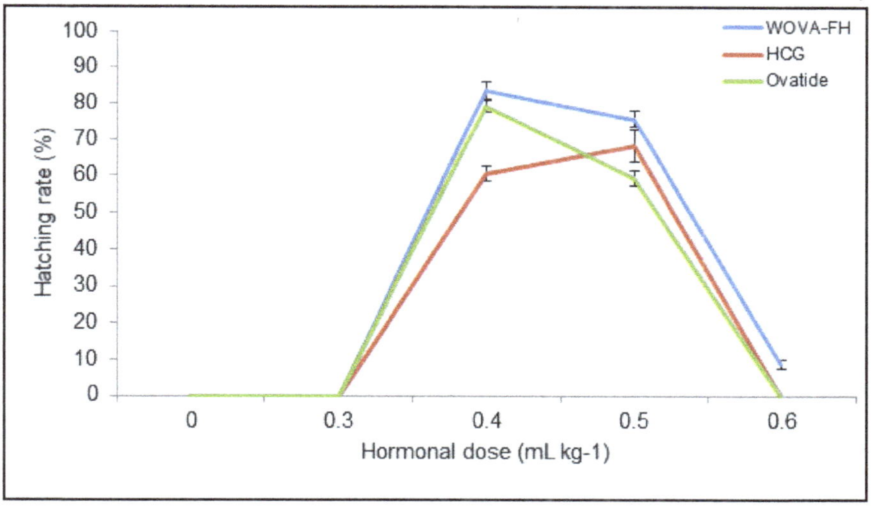

Fig. 8: Effect of different hormones and dosages on mean hatching rate (%) in *P. sophore*

Table 17: Description on embryonic development of *Puntius sophore* recorded during captive breeding experiment (Plate 50)

Developmental Stage	Descriptional characteristics	Mean Development time (Hours:min.)
Zygote period (1 cell)	Yolk-free cytoplasm begins to stream towards fish pole segregating blastodisc. Blastodiscs were not observable.	00:20
Cleavage Period (2 cells)	Two blastomers were rounded.	00:45

Developmental Stage	Descriptional characteristics	Mean Development time (Hours:min.)
Morula (64 cells)	Blastodisc consisted of many blastomers.	01:45
Blastula (256 cells)	Blastodermal size reduced and number of marginal cells increased.	02:45
Oblong stage	A smoothy outlined ellipsoidal shape was observed.	03:45
Sphere shape	Border between blastodisc and yolk flattened and spreaded gradually.	04:30
Dome shape	Blastoderm became dome like.	04:50
Gastrula period	Gastrulation begins forming epiboly.	05:30
30% epiboly	Blastoderm became flattened down onto the yolk sac and germ ring covered 30% of yolk sac.	05:35
50% epiboly	Germ ring covered 50% of yolk sac and neural plate started forming.	06:00
100% epiboly	Yolk invasion completed.	06:30
Bud formation	Epiboly came close to blastoderm completely. Tail bud appeared.	11:00
Segmentation period	First somite forms.	12:00
4 somites	4 somites visible with neurocoele cells in head.	14:00
8 somites	8 somites visible with optic vesicles and notochord.	16:00
12 somites	12 somites observable with prominent tail bud.	18:00
20 somites	20 somites visible. Heart beats were observable.	20:00
Pharyngula period	Eyes are clearly visible with bulgy yolk sac. Embryonic movement started.	25:45
	Pharyngula stage continued for another 14-16 hours.	-
Hatching	Fully transparent hatchlings was measured around 1.2-1.8 mm.	40:00
Newly hatched larvae	Head slightly bent on yolk. Responsive to stimulus.	

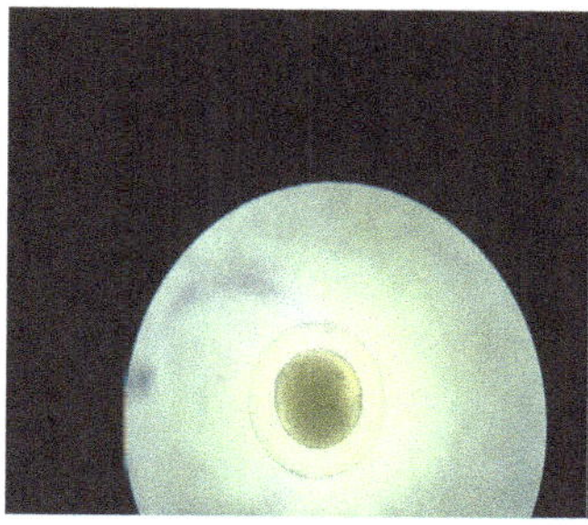

Fertilized egg

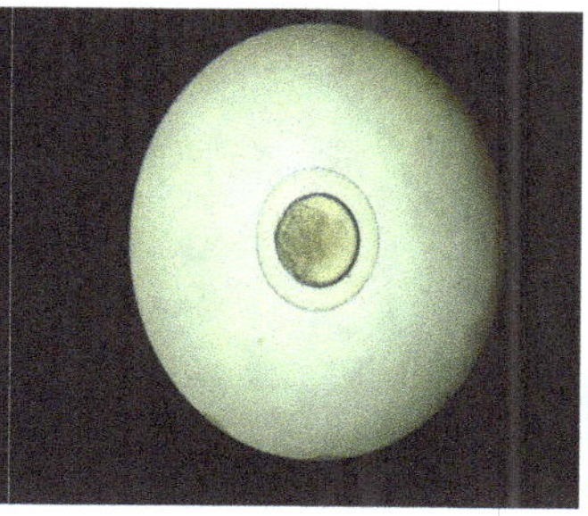

1-Cell stage

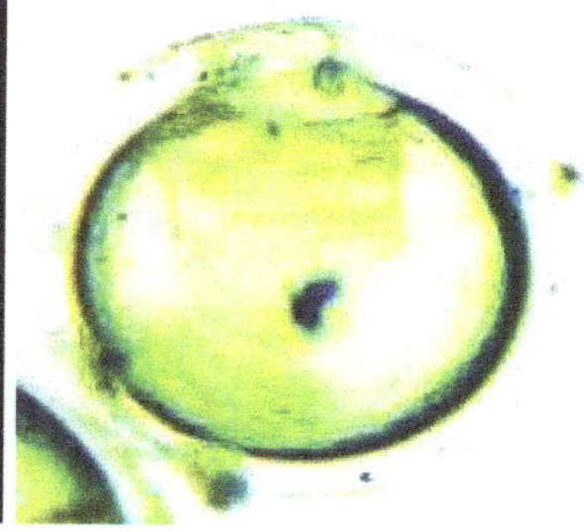

2-cell stage

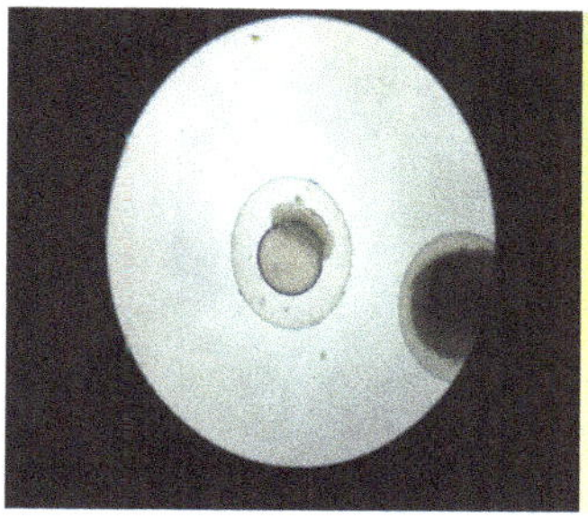

8-Cell stage

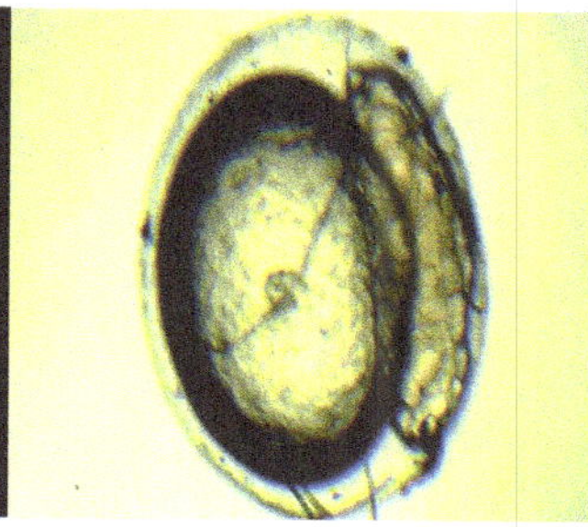

32-Cell stage

64-Cell stage

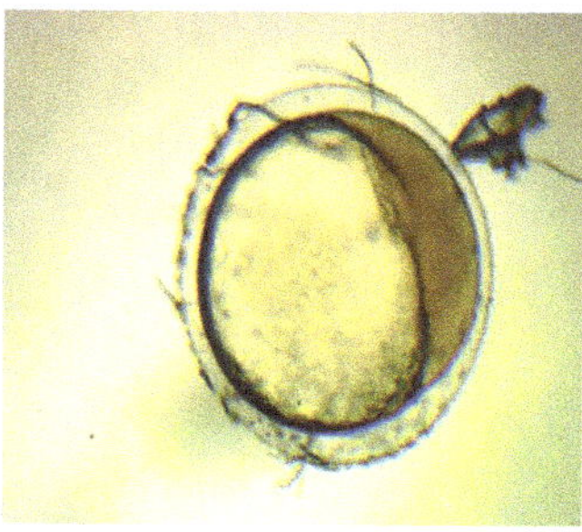

Morula stage

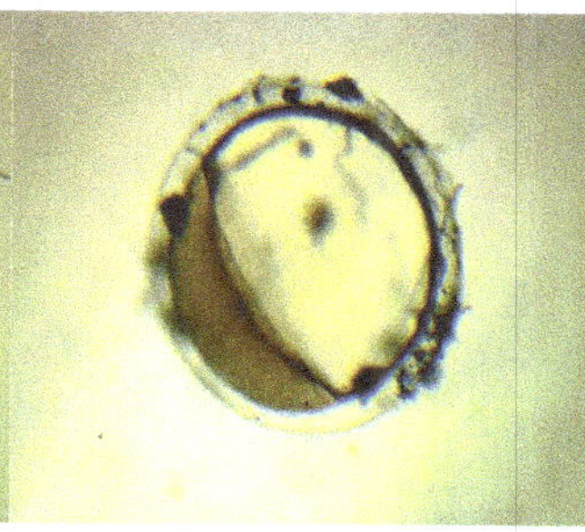

Blastula stage

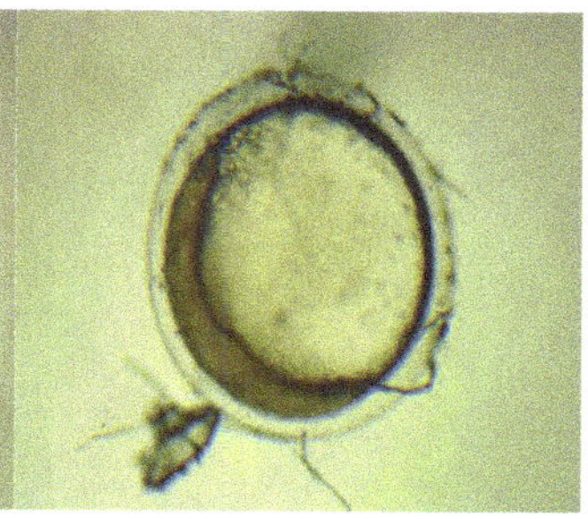

Gastrula stage

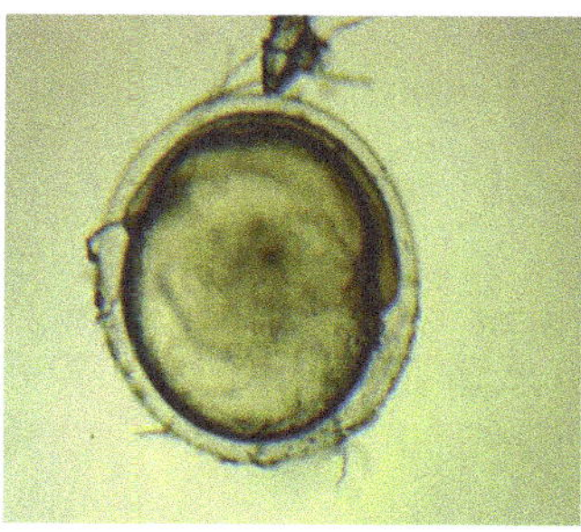

Embryonic stage

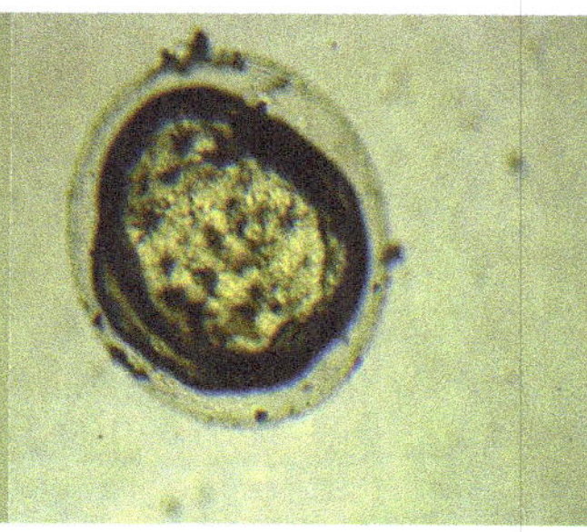

Epiboly stage

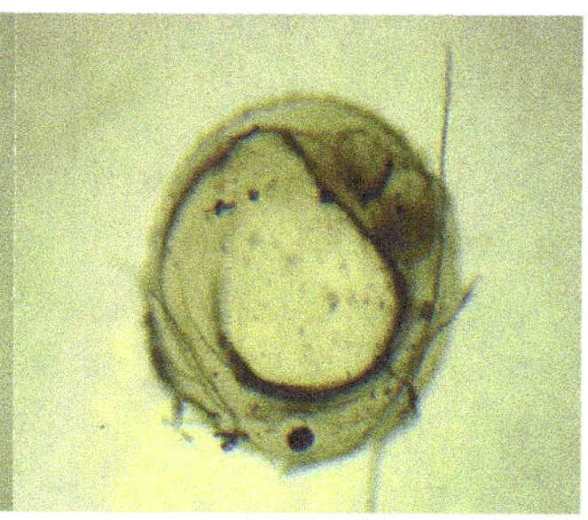

Somite formation stage

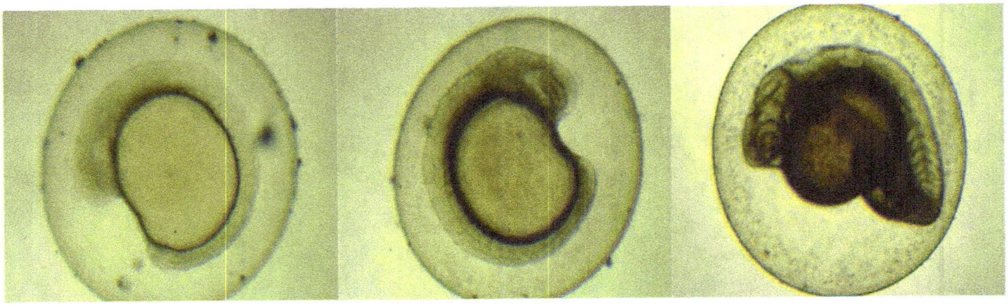

Somite formation (12-somites) Heart formation 15 Somite stage

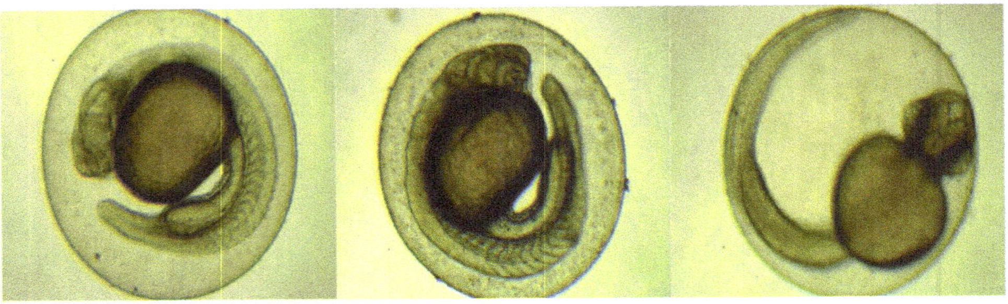

17 Somite stage 23- Somite stage Pharyngula stage

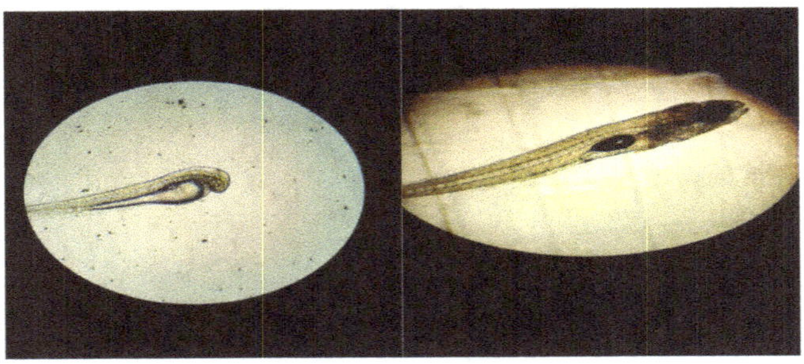

Larvae after hatching from chorion Larvae after yolk sac absorption

Plate 50: Embryonic developmental stages recorded during captive breeding of *P. sophore*

4.3 Mass production of Rotifer, *Brachionus calyciflorus* in outdoor culture system

Effect of different food type on rotifer growth and extent of cell concentration required for maximizing rotifer growth has been presented in Table 18 and 19. The data obtained was kept for normality testing of food types and feeding interval showed normal distribution ($p<0.05$).

4.3.1 Effect of different food types on the growth of Rotifer, *B. calyciflorus*

Different food types have showed significant difference ($p<0.05$) in the rotifer production. Higher rotifer count was observed in tanks fed with *C. vulgaris* (180.33±2.18 nos. mL^{-1}) on 8th day followed by Bakers yeast (132.00±9.03 nos. mL^{-1}), while the lowest growth was observed in tanks fed with *S. major* (72.33±4.09 nos. mL^{-1}) with initial population of 10.00±2.4 nos. mL^{-1} rotifers in each tank at 5% level of significance (Table 18; Fig. 9). Baker's yeast also gave good results (132.00±9.03 nos mL^{-1}) compared with other species which had less rotifers densities.

Table 18: Effect of different food types on Rotifer, *B. calyciflorus* production in outdoor culture system

Days	T_0	T_1	T_2	T_3	T_4
Day 0	10.00±2.1	10.00±1.5	10.00±1.4	10.00±2.1	10.00±1.1
Day 2	12.33±2.72^{a}	13.66±1.33^{a}	18.33±3.84^{a}	39.33±1.33^{a}	22.66±3.96^{a}
Day 4	21.00±8.88^{a}	25.33±3.13^{a}	39.00±2.64^{a}	101.33±6.35^{a}	90.66±5.66^{bc}
Day 6	15.00±1.73^{a}	41.66±7.96^{a}	82.67±10.80^{ab}	127.33±7.33^{b}	108.33±6.83^{bc}
Day 8	4.33±0.88^{a}	56.20±10.21^{a}	83.00±1.52^{ab}	180.33±2.18^{c}	125.33±5.36^{bc}
Day 10	1.00±0.57^{a}	58.00±3.64^{ab}	104.00±2.00^{b}	161.67±9.27^{bc}	132.00±9.03^{c}
Day 12	00^{a}	71.67±14.52^{ab}	101.33±7.21^{b}	166.67±2.84^{c}	104.00±2.00^{b}
Day 14	00^{a}	72.33±4.09^{ab}	115.00±10.00^{c}	172.67±8.48^{bc}	131.33±7.21^{bc}
p-value	0.650	0.012	0.002	0.023	0.021

Note: *Values with different superscripts in a row differs significantly at $p<0.05$ (n=3).*

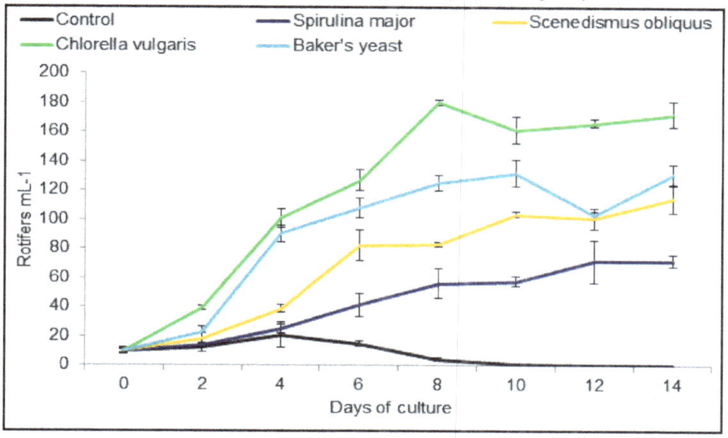

Fig. 9: Effect of different food type on Rotifer, *B. calyciflorus* production in outdoor culture system

4.3.2 Effect of feeding interval on rotifer, *B. calyciflorus* production

The food type efficiently preferred by rotifer was used at different intensity levels required to maximize rotifer production. Feeding interval at 2× and 3× day^{-1} had significant role on population growth of rotifers ($p<0.05$) (Table 19). Rotifers fed 3× day^{-1} with *C. vulgaris* showed significantly ($p<0.05$) higher population growth (256.33±3.8 nos. mL^{-1}) on 8th day compared with the tanks fed 2× day^{-1} (186.67±1.6 nos. mL^{-1}) and has been shown in Fig. 10.

Table 19: Effect of feeding interval (2× and 3× day^{-1}) on Rotifer production in outdoor culture system

Feeding Interval	2× day^{-1}	3× day^{-1}
Day 0	16.33±0.8^{a1}	18.00±1.5^{a1}
Day 2	32.00±3.6^{a1}	54.66±7.8^{a2}
Day 3	114.00±5.0^{a1}	184.00±12.0^{ab2}
Day 4	132.33±6.6^{ab1}	214.00±9.5^{b2}
Day 6	146.67±3.3^{b1}	246.00±6.5^{bc2}
Day 8	186.67±1.6^{c1}	256.33±3.8^{c2}
Day 10	171.00±8.0^{bc1}	231.67±8.8^{bc2}
Day 12	181.67±7.7^{bc1}	254.67±2.^{9bc2}
Day 16	198.34±7.4^{c1}	248.72±6.3^{bc2}
p-value	0.014	0.021

Note: *Mean values with different alphabets as superscripts in a column differs significantly (p<0.05); Mean values with different number as superscript in a row differs significantly (p<0.05).*

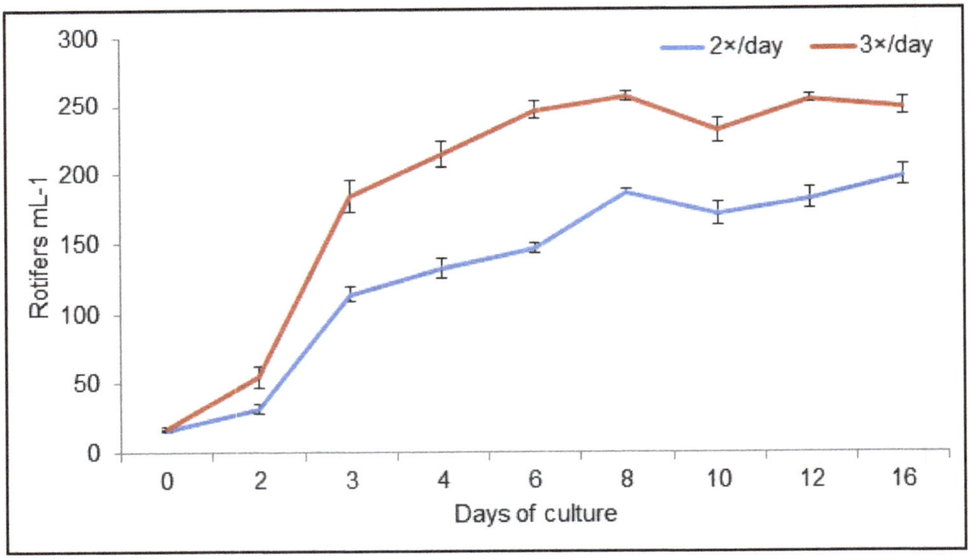

Fig. 10: Effect of feeding interval (2× and 3× day^{-1}) on Rotifer production in outdoor culture system

4.3.3 Water quality parameters

The range of physico-chemical parameters during culture of *B. calycifloruss* such as water temperature, dissolved oxygen, pH, NH_3 and others were well within the acceptable limits. During the experimental period the test water pH in 2× day^{-1} feeding tank ranged from 7.4 to 8.2 while dissolved oxygen ranged from 3.8 to 6.8 mg L^{-1}. The test water pH in 3× day^{-1} feeding tanks varied from 7.4-8.4 while dissolved oxygen from 3.2 to 7.0 mg L^{-1}. Ammonia was under acceptable limits and varied from 0.01 – 0.11 mg L^{-1} (Fig. 11). Conductivity ranged from 68.33 to 71.00 μM cm^{-1}, while total hardness and alkalinity recorded were 290.99 to 581.28 mg L^{-1} and 251.5 to 330.5 mg L^{-1} as $CaCO_3$, respectively (Fig. 12).

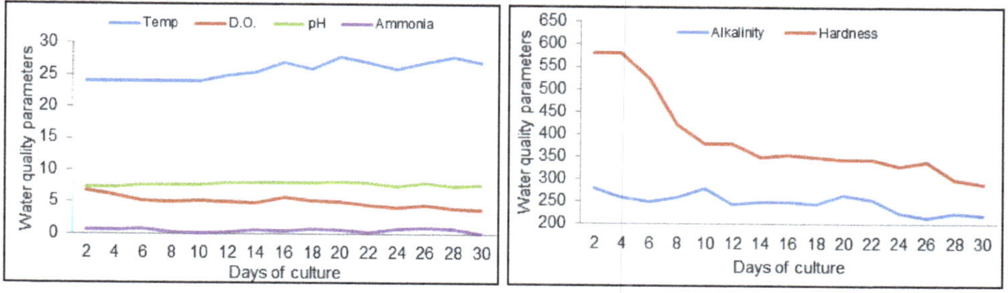

Fig. 11: Water quality parameters in 2× feeding intensity tanks

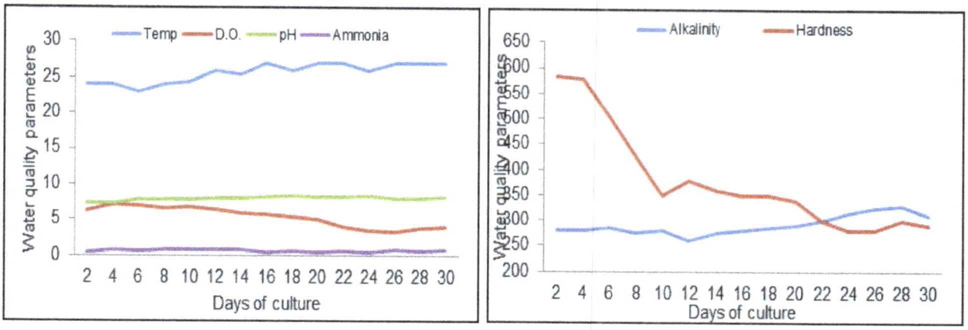

Fig. 12: Water quality parameters in 3× feeding interval tanks

4.4 Mass production of single cell algae, *Chlorella vulgaris*

Effect of light wavelengths on biomass production, specific growth rate, protein and pigment enhancement and cell size of *C. vulgaris* have been illustrated in Table 20 and 21. The data obtained on cell concentration and specific growth rate were kept for normality testing using different light wavelengths, showed normal distribution ($p<0.05$).

4.4.1 Effect of light wavelengths on *Chlorella vulgaris* production

Overall effect of different light wavelengths applied at 5 kLux has showed significance difference in biomass production of *C. vulgaris* within 6 days ($p<0.05$) (Table 20). The growth of *C. vulgaris* increased till 5th day in all sets and after, it showed stagnant growth as microalgae could have entered to its stationary phase. The highest cell count of *C. vulgaris* was observed in green light (65.220±0.17×10^6 cells mL^{-1}) followed by blue light (54.700±1.63×10^6 cells mL^{-1}) on 6th day (Fig. 13). The specific growth rate observed on 4th day was maximum in green light (0.79±0.90 d^{-1}) followed by blue light (0.70±0.70 d^{-1}). The lowest cell count (43.692±0.13×10^6 cells mL^{-1}) and growth rate (0.22±0.30 d^{-1}) of *C. vulgaris* was observed in red light compared with other sets at the end of experiment (Table 20 and Fig. 13).

Table 20: Effect of light wavelengths on cell concentration and specific growth rate of *Chlorella vulgaris*

		White	Red	Green	Blue	*p*-value
Cell concentration (Cell mL^{-1}) $\times 10^6$	Day 0	0.11±0.24[a]	0.10±0.11[a]	0.11±0.19[a]	0.11±0.16[a]	0.062
	Day 2	8.052±0.48[a]	7.739±0.66[a]	8.097±0.58[a]	7.671±0.58[a]	0.048
	Day 4	27.249±0.71[c]	25.715±0.79[c]	45.075±1.17[a]	37.185±0.11[b]	0.029
	Day 6	46.290±0.46[bc]	43.692±0.13[c]	65.220±0.17[a]	54.700±1.63[b]	0.043
Specific Growth Rate (day^{-1})	Day 2	0.30±0.07[b]	0.21±0.08[c]	0.51±0.05[a]	0.29±0.08[b]	0.033
	Day 4	0.29±0.3[c]	0.24±0.9[c]	0.79±0.3[a]	0.54±0.4[b]	0.019
	Day 6	0.51±0.3[bc]	0.22±0.3[c]	0.79±0.9[a]	0.70±0.7[b]	0.027

Note: *Values with different superscripts in a row differs significantly at $p<0.05$ (n=3)*

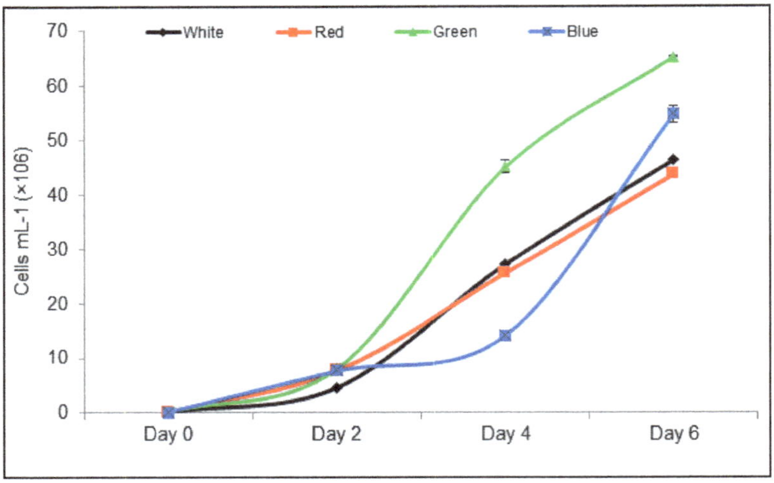

Fig. 13: Effect of different light wavelengths at 5k lux on *C. vulgaris* production in internal photobioreactor

4.4.2 Effect of different light wavelengths at 5k lux on cell size, protein and pigment enhancement in *C. vulgaris*

Light wavelength has played significant role on cell size, protein and pigment enhancement in *C. vulgaris* ($p<0.05$) and described in Table 21. The highest protein content was observed in the treatment illuminated with red light (0.132±0.14 mg mL^{-1}) followed by white light (0.049±0.01 mg mL^{-1}). The highest lipid content was observed in *C. vulgaris* grown under white light (0.252±0.24 mg mL^{-1}) followed by green light (0.210±0.31 mg mL^{-1}) at the same temperature. Chlorophyll content also varied significantly ($p<0.05$) with different light wavelengths. Highest content of Chlorophyll a (0.177±0.17 mg mL^{-1}) and chlorophyll b (0.141±0.13mg mL^{-1}) was observed in sets illuminated with green light followed by blue illumination (0.119±0.07 mg mL^{-1} and 0.122±0.02 mg mL^{-1}), respectively. On the other hand, highest astaxanthine production was observed in culture illuminated with red light (0.0126±0.08 mg mL^{-1}) followed by green light (0.0104±0.02 mg mL^{-1}). In addition, during cell count it was observed that the cell size of *C. vulgaris* has varied significantly ($p<0.05$). The superior cell size of *C. vulgris* was observed in the treatment illuminated with red light (5.2-9.4 μm) followed by white light (5.0-8.9 μm). The details of protein and pigments at different light wavelengths have been summarized in Fig. 14.

Table 21. Effect of different light wavelengths at 5k lux on cell size, protein and pigment enhancement of *C. vulgaris* in internal photobioreactor

	White	Red	Green	Blue	*p*-value
Protein content (mg mL^{-1})	0.049±0.01^{b4}	0.132±0.14^{c3}	0.025±0.05^{a}	0.012±0.04^{a}	0.022
Lipid content (mg mL^{-1})	0.252±0.24^{c5}	0.159±0.12^{a4}	0.210±0.31^{b}	0.116±0.17^{a}	0.0123
Chlorophyll a (mg mL^{-1})	0.085±0.24^{b3}	0.022±0.41^{a1}	0.177±0.17^{c}	0.119±0.07^{bc}	0.029
Chlorophyll b (mg mL^{-1})	0.076±0.05^{ab2}	0.031±0.02^{a2}	0.141±0.13^{c}	0.122±0.02^{bc}	0.0153
Astaxanthin (mg mL^{-1})	0.0058±0.07^{a1}	0.0126±0.08^{c5}	0.0104±0.02^{bc1}	0.0098±0.01^{b}	0.012
Cell size (μm)	5.0-8.9^{bc}	5.2-9.4^{c}	2.7-5.8^{a}	5.1-6.8^{b}	-

Note: *Mean values with different alphabet as superscripts in row differs significantly (p<0.05)*
Mean values with different number as superscripts in column differs significantly (p<0.05)

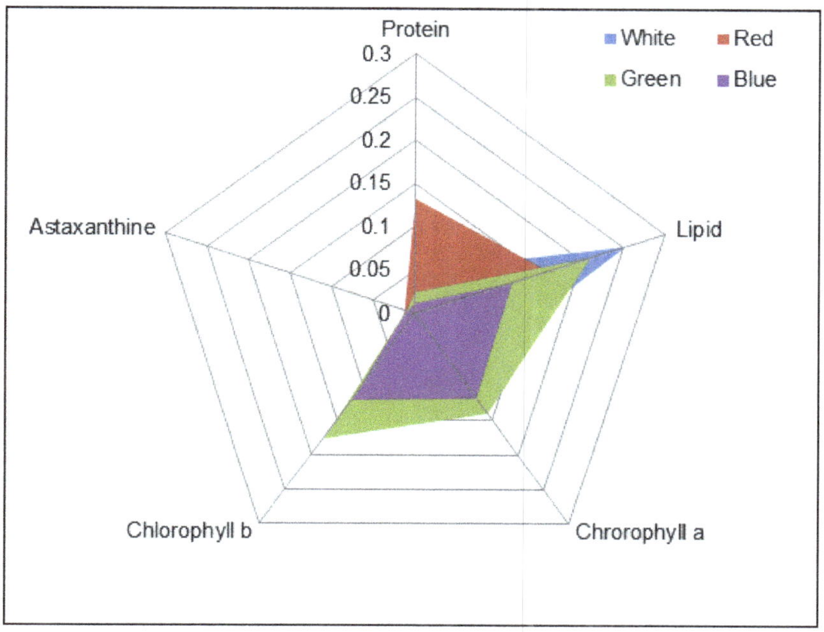

Fig. 14: Effect of different light wavelengths at 5k lux on protein and pigment enhancement in *C. vulgaris* produced using internal photobioreactor

4.4.3 Control of Ciliates

Ciliates posed serious problem in algal culture tanks throughout the culture period causing frequent crashes in algal culture tanks. The major ciliates observed in algal culture were *Lacrymaria* sp., *Paramecium* sp., *Coleps* sp. and *Dileptus* sp., are not desired as they compete for feed (Plate 51).

Common purified salt treatment at different dosage on the control of ciliates has played significant role over time during the culture of microalgae. No effect was observed even after 24 hours when salt was applied at 20 gm L^{-1} and 40 gm L^{-1}, but salt at 60 gm L^{-1} showed significant reduction in ciliates without affecting algal cells within 24 hours In contrast, salt at 80 gm L^{-1} showed that both ciliates and algal cells were alive for first 6 hours but algae became faint and after 15 hours both algal cells and ciliates were found dead (Table 22).

Table 22: Effect of common purified salt on control of Ciliates in algal, *C. vulgaris*, **culture tanks**

Salt (g) (NaCl)	6 hours	12 hours	18 hours	24 hours
20	Both algal cells and ciliates alive	Both alive	Both alive	Both alive
40	Both algal cells and ciliates alive	Both alive	Both alive	Both alive
60	Both algal cells and ciliates alive	Ciliates quantity decreased	Ciliates decreased	Ciliates dead. Algae alive
80	Both algal cells and ciliates alive	Both dead	Both dead	Both dead

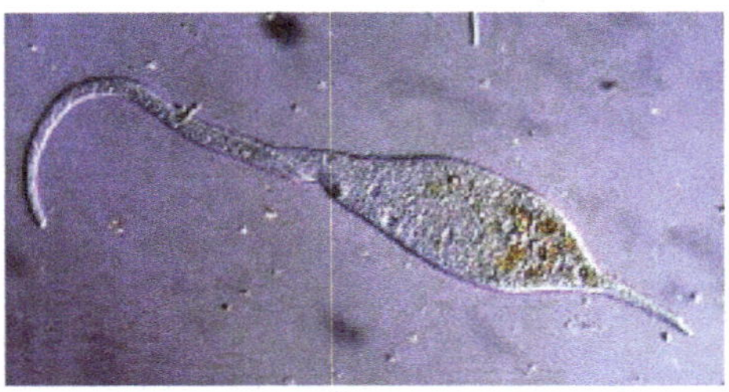

1. Lacrymaria

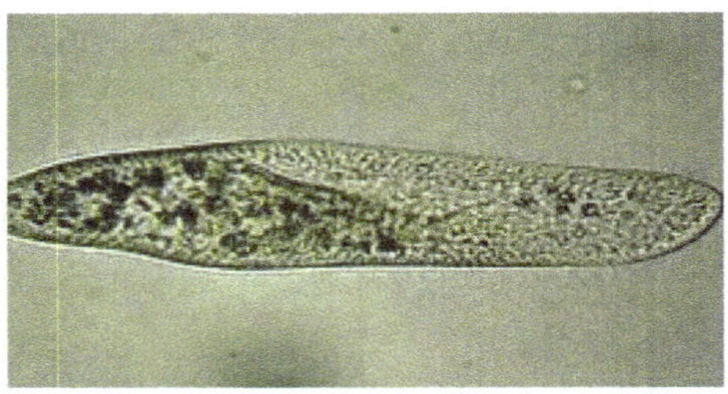

2. Paramecium

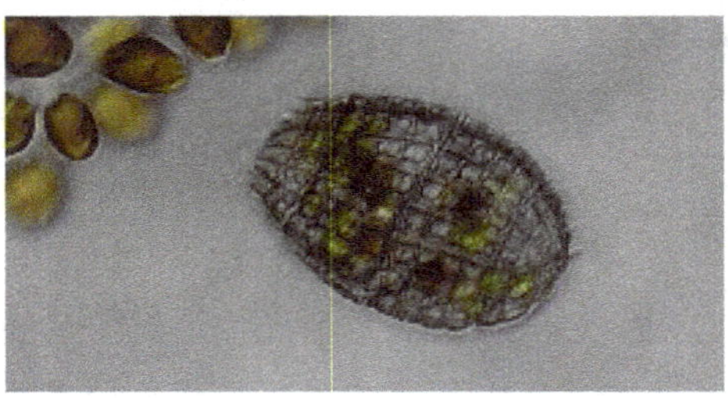

3. Coleps

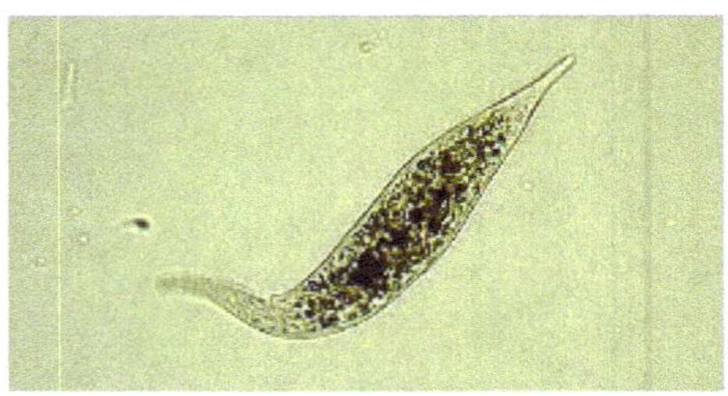

4. Dileptus

Plate 51: Major Ciliates observed in algal, *C. vulgaris*, culture tanks

4.5 Mass production of Rotifers, *B. calyciflorus* in indoor RAS based system

Effect of different food type on rotifer growth, determination of algal (*C. vulgaris*) consumption rate, effect of daily harvest on rotifer population and effect of RAS system on rotifer production has been illustrated in Table 23, 24, 25 and 26. The data obtained was kept for normality testing of different food types, minimum effective cell requirement by individual rotifer and daily harvest showed normal distribution ($p<0.05$).

4.5.1 Effect of different food type on rotifer production in indoor culture system

Overall effect of different food types on rotifer population was investigated and significant difference was observed among food types ($p<0.05$). The highest population growth was observed in tanks fed with *C. vulgaris* (365.33±2.18 nos. mL^{-1}) followed by Baker's yeast (245.33±5.36 nos. mL^{-1}) while the lowest growth was observed in tanks fed with *S. major* (86.00±10.21 nos. mL^{-1}) and *S. obliquus* (212.67±10.80 nos. mL^{-1}) on 8th day (Table 23) and then started declining from 10th day onwards (Fig. 15). At the end of experiment (30th day), rotifers from control group has vanished while tanks fed with *S. major* showed lowest count (57.44±6.28 nos. mL^{-1}) followed by *S. obliquus* (141.23±6.35 nos. mL^{-1}). On the other hand, tanks fed with *C. vulgaris* showed 229.74±6.32 nos. mL^{-1} average rotifer count followed by Baker's yeast (144.23±8.41 nos. mL^{-1}) (Fig. 15).

Table 23: Effect of different food type on Rotifer (*B. calyciflorus*) production in indoor culture system (Rotifers mL^{-1})

Days	Control	*Spirulina major*	*Scenedismus obliquus*	*Chlorella vulgaris*	Baker's Yeast
Day 0	10.00±2.1	10.00±1.5	10.00±1.4	10.00±2.1	10.00±1.1
Day 2	15.00±1.73^{a}	23.66±1.33^{a}	18.33±3.84^{a}	39.33±1.33^{b}	22.66±3.96^{a}
Day 4	27.00±8.88^{a}	54.66±7.96^{ab}	69.00±2.64^{ab}	131.33±6.35^{c}	90.66±5.66^{bc}
Day 6	12.33±2.72^{a}	75.33±3.13^{ab}	83.00±1.52^{b}	167.33±7.33^{c}	228.33±6.83^{c}
Day 8	4.33±0.88^{a}	86.00±10.21^{ab}	212.67±10.80^{c}	365.33±2.18^{d}	245.33±5.36^{c}
Day 10	1.00±0.57^{a}	92.33±4.09^{ab}	204.00±2.00^{c}	351.67±9.27^{d}	152.00±9.03^{bc}
Day 12	00^{a}	71.67±14.52^{b}	141.33±7.21^{bc}	346.67±2.84^{d}	104.00±2.00^{b}
Day 14	00^{a}	58.00±3.64^{a}	155.00±10.00^{bc}	362.67±8.48^{d}	141.33±7.21^{c}
Day 16	00^{a}	62.24±8.30^{a}	141.23±6.35^{bc}	241.61±9.31^{cd}	155.23±8.35^{bc}
Day 18	00^{a}	60.29±7.98^{a}	92.24±6.52^{b}	239.71±8.96^{cd}	141.32±9.68^{bc}
Day 20	00^{a}	55.33±6.93^{a}	104.11±6.48^{ab}	237.11±8.54^{cd}	92.24±5.36^{b}
Day 22	00^{a}	57.34±4.56^{a}	113.24±7.21^{ab}	228.61±7.96^{cd}	127.86±5.66^{c}

Days	Control	*Spirulina major*	*Scenedismus obliquus*	*Chlorella vulgaris*	Baker's Yeast
Day 24	00[a]	82.74±6.33[ab]	115.31±3.61[ab]	218.94±7.33[c]	131.74±7.71[bc]
Day 26	00[a]	90.56±9.20[ab]	79.78±5.15[b]	203.31±9.27[c]	117.96±7.49[bc]
Day 28	00[a]	69.26±4.78[a]	109.79±4.33[b]	211.66±2.64[c]	136.11±8.02[bc]
Day 30	00[a]	57.44±6.28[a]	139.88±3.32[ab]	229.74±6.32[cd]	144.23±8.41[bc]
p-value	0.000	0.002	0.001	0.001	0.002

Note: *Mean values with different alphabet as superscripts in rows differs significantly (p<0.05)*

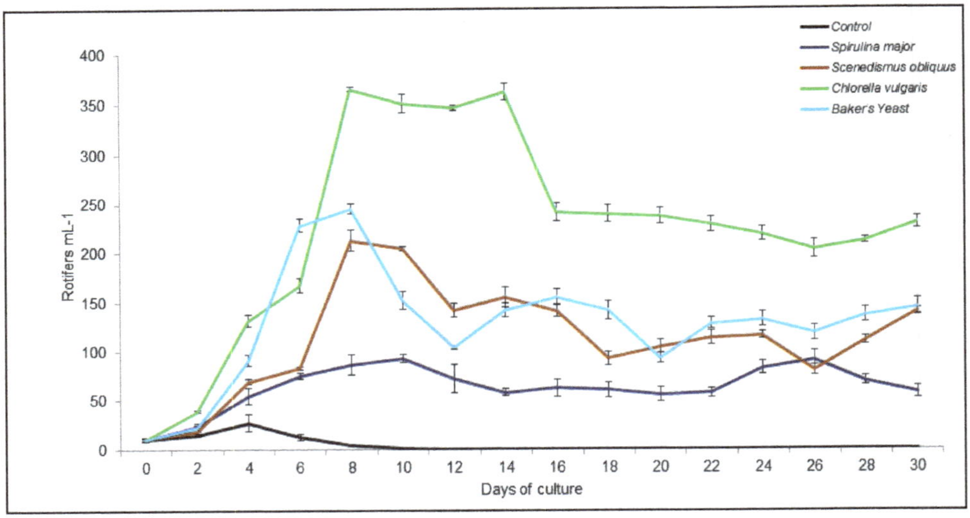

Fig. 15: Effect of different food type on Rotifer production in indoor culture system (Rotifers mL^{-1})

4.5.2 Estimation of algal, *Chlorella vulgaris*, consumption rate by individual rotifer, *Brachionus calyciflorus*

Algal cells, *C. vulgaris*, applied at different proportions to the same quantity of rotifers showed significant difference (*p<0.05*) on consumption and multiplication rate of rotifers. Almost all algal cells were consumed in T_3 (104010 cells Rotifer $^{-1}$) and T_4 (54930 cells Rotifer^{-1}) with the mean rotifer count of 0.48±0.02 nos. mL^{-1} and 0.08±0.03 nos. mL^{-1}, respectively, in 24 hours time period. Whereas, in T_1 (202620 cells Rotifer^{-1}) and T_2 (199010 cells Rotifer^{-1}) the quantum of rotifers recovered from jar was 1.08±0.09 nos. mL^{-1} and 1.08±0.04 nos. mL^{-1}, respectively, which was almost equal without showing significant difference (*p<0.05*). The highest quantity of rotifers was recovered from T_1 (544.63±44.03 nos. Jar^{-1}) followed by T_2 (541.33±21.18 nos. Jar^{-1}) while lowest was observed in control (7.33±17.38 nos. Jar $^{-1}$) (Table 24). Minimum algal cell requirement on the basis of algal consumption rate has been illustrated in Fig. 16.

Table 24: Estimation of algal cell, *Chlorella vulgaris*, consumption rate by Rotifers in Jar (500 mL)

	Rotifers Jar^{-1}	Algal cells Jar^{-1} (×1000)	Algal cells mL^{-1} (×1000)	Algal cells Rotifer^{-1} (×1000)	Algal cells recovered after 24 hours Jar^{-1} (×1000)	Algal cells consumed Rotifer^{-1} (×1000)	Rotifers recovered Jar^{-1}	Rotifers mL^{-1}
T_0	200 ± 5.4	0	0	0	0	0	7.33 ± 17.38^{a}	0.015 ± 0.04^{a}
T_1	200 ± 4.7	60524 ± 47.89	121.05 ± 0.95	202.62 ± 2.39	21036.70 ± 88.22^{a}	25.1 ± 9.41^{a}	544.63 ± 44.03^{b}	1.08 ± 0.09^{b}
T_2	200 ± 4.2	39892 ± 63.98	79.60 ± 1.27	199.01 ± 3.20	2416.70 ± 65.01^{a}	22.08 ± 0.33^{ab}	541.33 ± 21.18^{b}	1.08 ± 0.04^{b}
T_3	200 ± 3.1	20803 ± 14.33	41.61 ± 2.87	104.01 ± 7.17	981.10 ± 75.97^{b}	14.91 ± 0.38^{b}	240.61 ± 11.60^{c}	0.48 ± 0.02^{c}
T_4	200 ± 3.2	10987 ± 73.61	21.97 ± 1.47	54.93 ± 3.68	375.63 ± 10.43^{c}	7.88 ± 0.90^{c}	43.33 ± 14.49^{a}	0.08 ± 0.03^{a}
p-	-	-	-	-	0.025	0.031	0.043	0.014

Note: *Mean values with different alphabets as superscripts in a column differs significantly ($p<0.05$).*

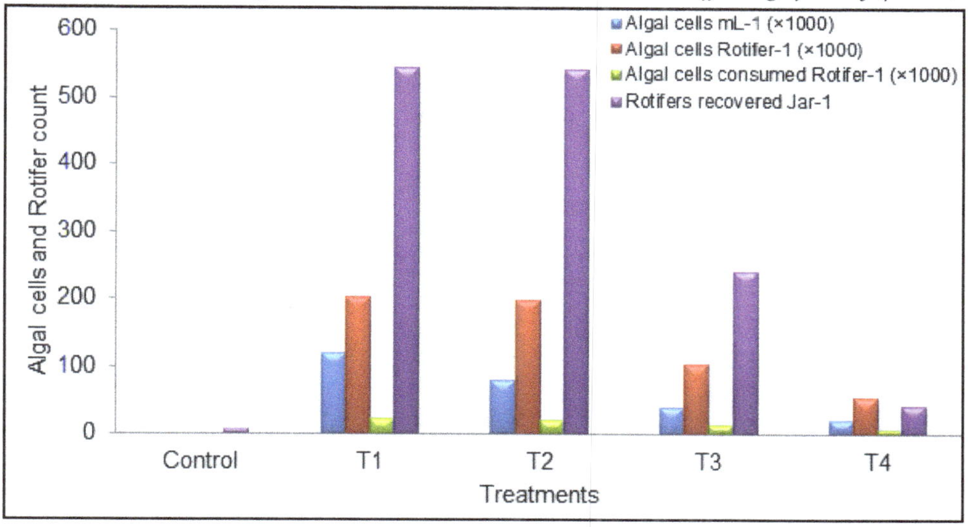

Fig. 16: Algal cell, *Chlorella vulgaris,* consumption rate by Rotifers

4.5.3 Effect of daily harvest rate on Rotifer production:

Daily harvest rate showed significant ($p<0.05$) role on rotifer production during experiment. Initial rotifer quantity was maintained at an average of 250±18.20 nos. mL^{-1}. The highest harvested quantity of rotifers was observed at 70% harvest rate (9100±19.75×10^{5} nos.) followed by 50% (6275±29.80×10^{5} nos.) on 2nd day but was decreased gradually to 490±48.14×10^{5}and 2050±31.26×10^{5} nos., respectively, on 8th day. When tanks were harvested at 30% of the total volume, almost equal quantity of rotifers was observed throughout the experiment (381±19.45 nos. mL^{-1}). At 10% harvest rate, comparatively, less (157±29.80 nos. mL^{-1}), count was observed while the lowest count was observed in 70% harvest rate (11±9.37 nos. mL^{-1}) in 8 days culture period (Table 25). Daily harvest rate at 30% has showed stable quantity of rotifers throughout the experiment (Fig. 17).

Table 25: Effect of daily harvest on Rotifer production in Recirculatory Aquaculture System (RAS)

	Rotifer/ mL	Harvested Rotifers x 1E+05	Rotifers mL^{-1} after 2 Days	Harvested Rotifers x 1E+05	Rotifers mL^{-1} after 4 Days	Harvested Rotifers x 1E+05	Rotifers mL^{-1} after 6 Days	Harvested Rotifers x 1E+05	Rotifers mL^{-1} after 8 Days
T_c	234 ± 12.30	0	263 ± 14.78^{bc1}	0	198 ± 9.65^{c2}	0	162 ± 29.00^{bc2}	0	98 ± 18.55^{c3}
T_1	257 ± 18.20	1286 ± 15.87^{d3}	284 ± 18.79^{b1}	1420 ± 18.55^{d2}	260 ± 10.48^{b1}	1300 ± 11.20^{d2}	182 ± 16.25^{b2}	4100 ± 23.15^{b1}	157 ± 29.80^{b3}
T_2	248 ± 18.24	3720 ± 14.28^{c4}	319 ± 25.24^{a1}	4745 ± 23.69^{c3}	356 ± 28.13^{a1}	5340 ± 15.11^{a2}	395 ± 11.12^{a2}	5925 ± 35.14^{a1}	381 ± 19.45^{a3}
T_3	251 ± 21.00	6275 ± 29.80^{b2}	296 ± 16.56^{b1}	7400 ± 25.29^{a1}	209 ± 16.13^{c2}	5220 ± 29.41^{b2}	82 ± 24.42^{c3}	2050 ± 31.26^{c4}	82 ± 11.20^{c3}
T_4	260 ± 19.45	9100 ± 19.75^{a1}	204 ± 21.11^{d1}	7140 ± 13.84^{b2}	64 ± 9.37^{d2}	2240 ± 14.57^{c3}	14 ± 7.96^{d3}	490 ± 48.14^{d4}	11 ± 9.37^{d3}

Note: *Mean values with different alphabets as superscripts in column differs significantly ($p<0.05$)*
Mean values with different numbers as superscripts in row differs significantly ($p<0.05$)

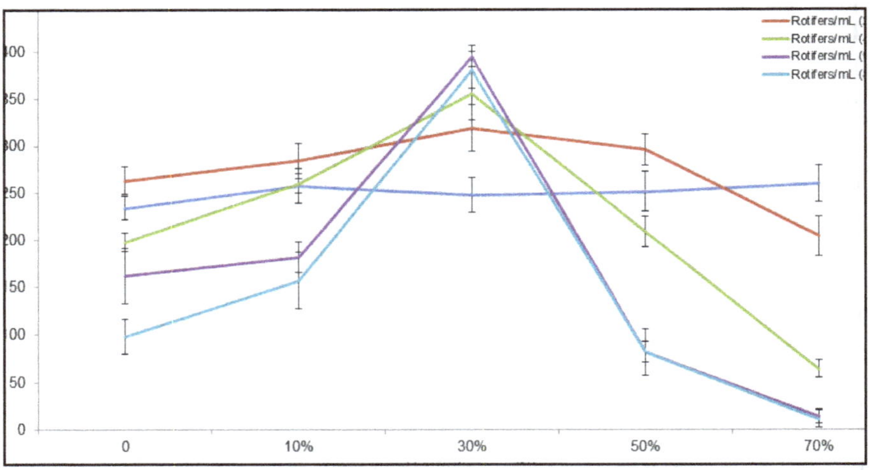

Fig. 17: Effect of daily harvesting rate on Rotifer production in Recirculatory Aquaculture System (RAS)

4.5.4 Effect of Recirculatory Aquaculture system (RAS) on Rotifer production

Recirculatory system has played significant role on rotifer production by maintaining water quality throughout the culture period, except temperature. The highest rotifer production was observed during summer (394.12±15.41 nos. mL^{-1}) but was started declining gradually during monsoon (349.57±25.44 nos. mL^{-1}) followed by winter (301.21±14.33 nos. mL^{-1}) and again reached to its normal stage during summer (404.43±24.33 nos. mL^{-1}) (Fig. 18). Water quality parameters during the culture period were remained within the acceptable limits. The ranges of different water quality parameters during different seasons were depicted in Table 26. The dissolved oxygen, pH, ammonia, alkalinity, hardness and temperature during culture period ranged from 3.8-6.8 mg L^{-1}, 7.4-8.4, 0.1-1.1 mg L^{-1}, 260-330 mg L^{-1}, 280-650 mg L^{-1} and 24.0-32.9°C, respectively. Variation of environmental parameters has been depicted in Fig. 19, 20 and 21.

Table 26: Water quality parameter and rotifer numbers (range) observed during study period

Parameters	Summer	Monsoon	Winter	Summer	Monsoon
Temperature (°C)	27.6-31.4	24.0-31.4	24.2-30.1	27.0-32.9	24.0-31.1
D.O. (mg L^{-1})	3.8-6.8	4.0-5.4	4.0-5.4	4.0-5.8	3.8-5.8
pH	7.4-8.4	7.6-8.2	7.4-8.4	7.4-8.4	7.6-8.4
Ammonia (mg L^{-1})	0.1-1.1	0.1-1.0	0.2-1.0	0.3-1.1	0.1-0.8
Alkalinity (mg L^{-1})	260-330	260-224	260-330	280-320	260-320
Hardness (mg L^{-1})	280-650	280-650	420-520	300-650	300-650
Rotifer quantity (nos. mL^{-1})	384-406	301-396	254-403	398-415	370-400

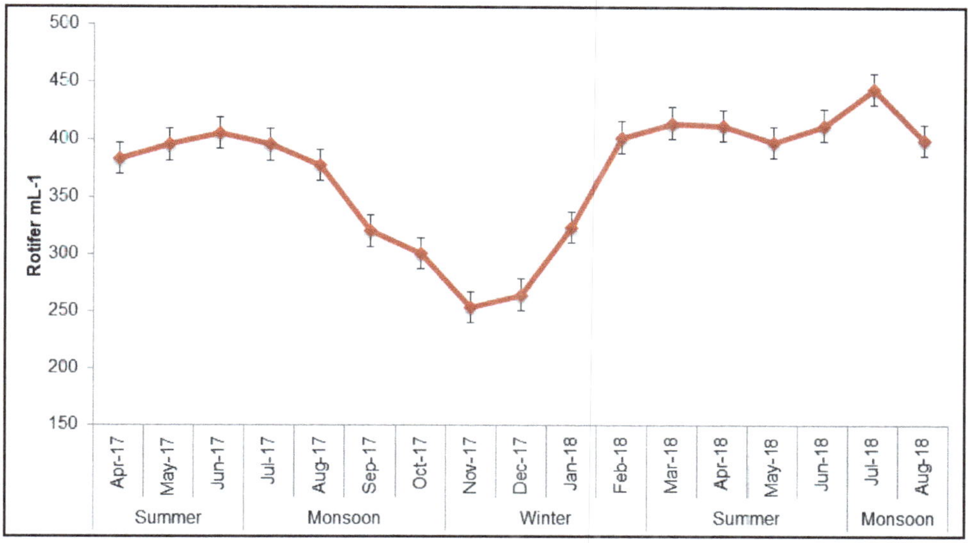

Fig. 18: Monthly Rotifer production in RAS system.

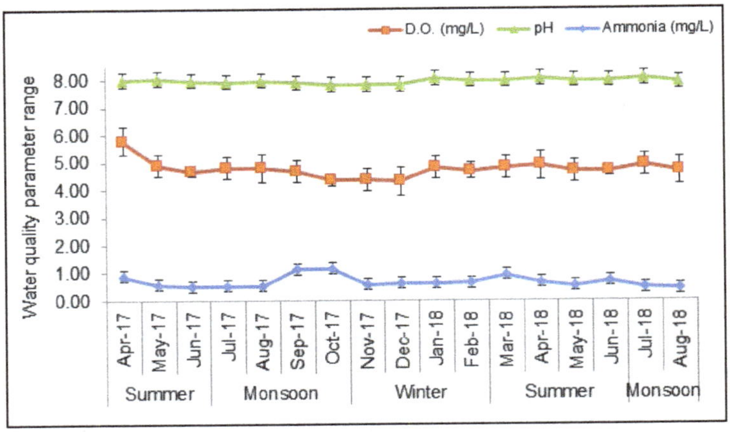

Fig. 19: Dissolved oxygen, pH and ammonia during culture period

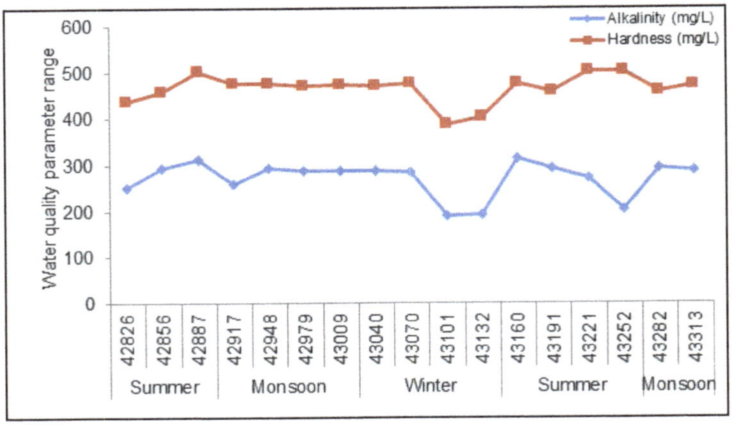

Fig. 20: Alkalinity and Hardness during culture period

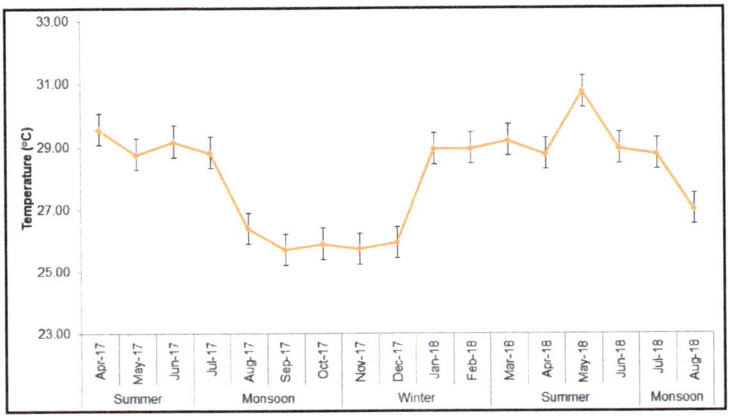

Fig. 21: Temperature variation during culture period of rotifer in RAS system

4.6 Formulation of probiotic incorporated encapsulated feed

4.6.1 Isolation, identification and characterization of *Lactobacillus sp.* from fish intestine

Totally, 11 colony numbers were observed in intestine of *D. dangila* while 09 colonies were observed in *P. sophore*. Total plate counts of bacteria in fish intestine were determined on TSA medium and showed 1.7×10^7 CFU g^{-1} population in intestine of *D. dangila* and around 2.1×10^7 CFU g^{-1} in intestine of *P. sophore*.

4.6.1.1 Morphological analysis of *Lactobacillus sp.* from fish intestine

The results of colony morphology, motility and appearance have been described in Table 27 and 28. The colonies from *D. dangila* were almost coccid rod shaped (06 colonies) appearing yellow to yellowish white lacking motility. The coccid shaped colonies (04 colonies) were cream white to yellow white in appearance and showed motility. Whereas, unicellular rod structured colony (colony 10) was golden yellow and motile (Plate 36).

On the other hand, the coccid rod shaped (05) colonies from *P. sophore* were seemed to be yellow. The coccid shaped colonies (04) were cream white in appearance and showed motility. Whereas, unicellular rod structured colony (colony 10) was golden yellow and motile (Plate 37).

Table 27: Colony morphology of isolated bacteria (11 colonies) from *D. dangila* intestine (Plate 36 A and B).

Colony number	Colony Morphology	Motility	Colony Appearance
Colony 1	Coccid rod	Non-motile	Yellowish white
Colony 2	Coccid rod	Non-motile	Yellow
Colony 3	Coccid rod	Non-motile	Yellow
Colony 4	Coccid rod	Non-motile	Yellow
Colony 5	Coccid	Motile	Cream white
Colony 6	Coccid rod	Non-motile	White
Colony 7	Coccid	Motile	Cream white
Colony 8	Coccid	Motile	Cream white
Colony 9	Coccid	Non-motile	Yellow white
Colony 10	Unicellular rod	Motile	Golden yellow
Colony 11	Coccid rod	Non-motile	Yellow

Table 28: Colony morphology of isolated bacteria (09 colony) from *P. sophore* (Plate 37 A and B)

Colony number	Colony Morphology	Motility	Colony Appearance
Colony 1	Coccid rod	Non-motile	Yellow

Colony number	Colony Morphology	Motility	Colony Appearance
Colony 2	Coccid rod	Non-motile	Yellow
Colony 3	Coccid	Motile	Cream white
Colony 4	Coccid rod	Non-motile	White
Colony 5	Coccid	Motile	Cream white
Colony 6	Coccid	Motile	Cream white
Colony 7	Coccid	Non-motile	Yellow
Colony 8	Unicellular rod	Motile	Golden yellow
Colony 9	Coccid rod	Non-motile	Yellow

4.6.1.2 Biochemical analysis of *Lactobacillus sp.* from fish intestine

The isolated colonies from both the species were kept for different biochemical test for confirmation of lactic acid bacteria. The results of biochemical analysis showed that the isolated colonies from the intestine of *D. dangila* were Gram's positive, except one (colony 10). The antagonistic effect of colonies in resistance test to *Aeromonas hydrophila* showed either no zone formation or the zone less than 3 mm in all colonies except colony 7, 8 and 10 which showed agglutination with zones less than 3 mm. The oxidative fermentation test showed fermentative reaction in all colonies except 7 and 8 which resulted in oxidative reaction, but colony 6 showed non saccharolytic reaction. The catalase and indole test showed negative reaction in almost all colonies except indole test in colony 8 which showed positive reaction (Table 29).

The results of biochemical analysis showed that the isolated colonies from the intestine of *P. sophore* were Gram's positive, except one (colony 8). The antagonistic effect of colonies in *Aeromonas hydrophila* resistance test showed either no zone formation or the zone less than 3 mm in all colonies except colony 6, 7 and 8 which showed agglutination with zones less than 3 mm. The oxidative fermentation test showed fermentative reaction in all colonies except 6 and 7 which resulted in oxidative reaction, but colony 5 showed non saccharolytic reaction. The catalase and indole test showed negative reaction in almost all colonies, except indole test in colony 7 which showed positive reaction (Table 30).

The susceptibility and resistance pattern obtained with three isolates from each of *D. dangila* and *P. sophore* has been described in Table 31. Observation specified that the isolates from both the species were less susceptible to antibiotics (no zone or zone less than 10 mm). In case of tetracycline, chloramphenicol, ofloxacin and ciprofloxacin all isolates from both the species showed zones less than 10 mm resulting lower resistance while in case of erythromycin, isolates from *D. dangila* showed less resistance (zones <10 mm) compared with *P. sophore* isolates. On the other hand, isolates from both the species showed more resistance against ampicilin, ceftazidime and penicillin (Table 31 and Plate 38).

Table 29: Biochemical tests of isolated colonies from *D. dangila* intestine

Colony	Gram's test	*Aeromonas hydrophila* resistance test	Oxidative Fermentation test	Indole test	Catalase test
1	Positive	Zone less than 3 mm	Fermentative	Negative	Negative
2	Positive	Zone less than 3 mm	Fermentative	Negative	Negative
3	Positive	No zone	Fermentative	Negative	Negative
4	Positive	Zone less than 3 mm	Fermentative	Negative	Negative
5	Positive	Zone less than 3 mm	Fermentative	Negative	Negative
6	Positive	No zone	Non Saccharolytic	Negative	Negative
7	Positive	Agglutination and zone less than 3 mm	Oxidative	Negative	Negative
8	Positive	Agglutination and zone less than 3 mm	Oxidative	Positive	Negative
9	Positive	Zone less than 3 mm	Fermentative	Negative	Negative
10	Negative	Agglutination and zone less than 3 mm	Fermentative	Negative	Negative
11	Positive	No zone	Fermentative	Negative	Negative

Table 30: Biochemical tests of isolated colonies from *P. sophore* intestine

Colony	Gram's test	*Aeromonas hydrophila* resistance test	Oxidative Fermentation test	Indole test	Catalase test
1	Positive	Less than 3 mm	Fermentative	Negative	Negative
2	Positive	No zone	Fermentative	Negative	Negative
3	Positive	Less than 3 mm	Fermentative	Negative	Negative
4	Positive	Less than 3 mm	Fermentative	Negative	Negative
5	Positive	No zone	Non-Saccharolytic	Negative	Negative
6	Positive	Agglutination and zone less than 3 mm	Oxidative	Negative	Negative
7	Positive	Agglutination and zone less than 3 mm	Oxidative	Positive	Negative
8	Negative	Agglutination and Less than 3 mm	Fermentative	Negative	Negative
9	Positive	No zone	Fermentative	Negative	Negative

Table 31: Antibiotic susceptibility test of isolates from *D. dangila* **and** *P. sophore* **(Plate 38 A and B)**

Danio dangila									
		Tetracycline	Penicillin	Erythromcin	Chloramphenicol	Ofloxacin	Ciprofloxacin	Ampicillin	Ceftazidime
(µg/disc)		30	10	10	10	10	10	10	30
Isolate number	2	<10 mm	No zones	<10 mm	<10 mm	<10 mm	<10 mm	No zones	No zones
	6	<10 mm	No zones	<10 mm	<10 mm	<10 mm	<10 mm	No zones	No zones
	11	< 10 mm	No zones	<10 mm	<10 mm	<10 mm	<10 mm	No zones	No zones
Puntius sophore									
Isolate number	2	<10 mm	No zones	No zones	<10 mm	<10 mm	<10 mm	No zones	No zones
	5	<10 mm	No zones	No zones	<10 mm	<10 mm	<10 mm	No zones	No zones
	9	<10 mm	<10 mm	No zones	<10 mm	<10 mm	<10 mm	No zones	<10 mm

Note: *Zones with less than 10 mm are less resistant.*

No-zone formations are more resistant.

Statistical analysis proved that the growth rate of *Lactobacillus sp.* varied significantly ($p<0.05$) with pH level and bile salt concentration observed at optical density of 600 nm (Table 29). Results substantiated that the growth of *Lactobacillus sp.* has been reduced considerably at low pH (3) for both *D. dangila* (0.0042±0.0002 cfu mL^{-1}) and *P. sophore* (0.0039±0.0004 cfu mL^{-1}). But effective growth was observed from pH 4 onwards and showed highest growth at neutral pH (7) for both *D. dangila* (0.0160±0.0006 cfu mL^{-1}) and *P. sophore* (0.0162±0.0007 cfu mL^{-1}).

On the other hand, three concentrations (0%, 0.15% and 0.30%) of bile salt showed viability in both the species (Table 29). Significantly higher growth of *Lactobacillus sp.* was observed at 0.01% bile salt concentration in both *D. dangila* (0.1250±0.0021 cfu mL^{-1}) and *P. sophore* (0.1290±0.0014 cfu mL^{-1}). The growth rate was reduced considerably as bile salt concentration increased. The lowest growth was observed at 0.3% bile salt concentration in both *D. dangila* (0.0183±0.0012 cfu mL^{-1}) and *P. sophore* (0.0163±0.0008 cfu mL^{-1}).

Table 32: pH and bile salt tolerance for *D. dangila* and *P. sophore* isolates

	pH tolerance (cfu mL^{-1})	
pH	***D. dangila***	***P. sophore***
3	0.0042±0.0002	0.0039±0.0004
4	0.0081±0.0001	0.0078±0.0004
5	0.0126±0.0002	0.0132±0.0001
6	0.0159±0.0002	0.0161±0.0003
7	0.0160±0.0006	0.0162±0.0007
8	0.0143±0.0005	0.0148±0.0004
9	0.0122±0.0002	0.0123±0.0002
p-value	0.000	0.000
Bile salt (%)	Bile salt (%) tolerance (cfu mL^{-1})	
0.01	0.1250±0.0021	0.1290±0.0014
0.15	0.0811±0.0031	0.0731±0.0017
0.3	0.0183±0.0012	0.0163±0.0008
p-value	0.000	0.000

4.6.2 Preparation and characterization of probiotic incorporated encapsulated feed

LAB viability test has showed no significant loss of *Lactobacillus sp.* after 24 hours of encapsulation. The loss of viability of lactic acid bacteria (LAB) above 15% was observed in feed stored at room temperature. Subsequently, samples stored at -20°C showed high viability of LAB with less than 08% of loss (Table 33, Plate 42). The proximate composition of feed has been explained in section 3.6.1.

Table 33: LAB viability test after encapsulation (24 hours).

Initial CFU g^{-1}	Final CFU g^{-1}		Viability (%)	
	At room temperature	At -20°C	At room temperature	At -20°C
1×10^{1}	0.81×10^{1c}	0.92×10^{1c}	85.1 ± 0.15^{c}	92.3 ± 0.12^{c}
1×10^{3}	8.421×10^{2b}	9.331×10^{2b}	84.2 ± 0.13^{b}	93.3 ± 0.13^{b}
1×10^{5}	8.578×10^{4a}	9.598×10^{4a}	85.78 ± 0.18^{a}	95.98 ± 0.11^{a}
1×10^{7}	8.421×10^{6b}	9.534×10^{6a}	84.21 ± 0.18^{b}	95.34 ± 0.14^{a}
P-value	0.000	0.001	0.000	0.001

Note: *Values with different alphabets as a superscripts in a column differs significantly at $p<0.05$ (n=3)*

Values with different numbers as a superscripts in a row differs significantly at $p<0.05$ (n=3)

4.6.3 Effect of probiotic incorporated feed on larval growth

Probiotic incorporated encapsulated feed with different concentrations of *Lactobacillus sp.* (10^1 cfu g^{-1}, 10^3 cfu g^{-1}, 10^5 cfu g^{-1} and 10^7 cfu g^{-1}) has showed significant ($p<0.05$) effect on larval growth, specific growth rate, average daily gain, feed conversion ratio and survival rate of both *D. dangila* and *P. sophore* (Table 34a and 34b). Treatment 3 (T_3) has showed highest growth in both *D. dangila* (1.0233±0.033 g) and *P. sophore* (0.8233±0.003 g) followed by T_2 (0.9800±0.030 g; 0.8133±0.007 g), respectively. While, the lowest growth was observed in T_4 for *D. dangila* (0.8733±0.048 g) and *P. sophore* (0.7467±0.023 g), respectively. The highest specific growth rate was observed in T_2 for *D. dangila* (0.1324±0.0023%) followed by T_3 (0.1307±0.0021%) but in *P. sophore* (0.1384±0.0015%) it was observed in T_3 (0.1384±0.0015%) followed by T_2 (0.1368±0.0011%), whereas, the lowest specific growth rate was observed in T_4 for both *D. dangila* (0.1258±0.0013%) and *P. sophore* (0.1190±0.0030%), respectively. The highest average daily gain was observed in T_3 for *D. dangila* (0.0334±0.0001 g) and *P. sophore* (0.0270±0.0001 g) followed by T_2 (0.0267±0.0002 g; 0.0320±0.0010 g) respectively, while the lowest detected in T_4 (0.0284±0.0013 g; 0.0242±0.0008 g). Consequently, in T_3, the highest survival rate (78.41±0.84%; 90.08±0.93%) with efficient feed conversion ratio (1.51±0.19; 1.54±0.02) was observed for *D. dangila* and *P. sophore,* respectively, followed by T_2. The lowest survival rate for *D. dangila* and *P. sophore* was seen in T_0 (70.70±1.37%) and T_4 (77.03±2.52%), respectively. The highest feed conversion ratio was seen in T_4 for both *D. dangila* (2.15±0.26) and for *P. sophore* (2.21±0.03).

Table 34a: Effect of probiotic incorporated feed on growth indices of *D. dangila* larvae

Probiotic feed (cfu g^{-1})	Final weight (g)	Specific growth rate (%)	Average daily gain (g)	Feed conversion ratio	Survival rate (%)
Danio dangila					
T_0	0.9733 ± 0.024^{ab}	0.1301 ± 0.0013^{b}	0.0318 ± 0.0002^{b}	2.07 ± 0.02^{c}	70.70 ± 1.37^{c}
T_1	0.9767 ± 0.043^{a}	0.1324 ± 0.0023^{a}	0.0319 ± 0.0021^{ab}	1.91 ± 0.02^{b}	77.66 ± 0.39^{a}
T_2	0.9800 ± 0.030^{ab}	0.1286 ± 0.0011^{b}	0.0320 ± 0.0010^{a}	1.85 ± 0.02^{b}	77.59 ± 0.67^{b}
T_3	1.0233 ± 0.033^{a}	0.1307 ± 0.0021^{a}	0.0334 ± 0.0001^{a}	1.51 ± 0.19^{a}	78.41 ± 0.84^{a}
T_4	0.8733 ± 0.048^{b}	0.1258 ± 0.0013^{c}	0.0284 ± 0.0013^{c}	2.15 ± 0.26^{c}	72.30 ± 0.97^{c}
p-value	0.021	0.014	0.019	0.005	0.005

Note: *Mean values with different superscripts in a column differs significantly at p<0.05 (n=3)*

Table 34b: Effect of probiotic incorporated feed on growth indices of *P. sophore* larvae

Probiotic feed (cfu g^{-1})	Final weight (g)	Specific growth rate (%)	Average daily gain (g)	Feed conversion ratio	Survival rate (%)
Puntius sophore					
T_0	0.8067 ± 0.003^{b}	0.1360 ± 0.0009^{b}	0.0264 ± 0.0001^{b}	2.16 ± 0.03^{c}	78.36 ± 1.67^{c}
T_1	0.8033 ± 0.186^{b}	0.1362 ± 0.0011^{b}	0.0264 ± 0.0006^{b}	1.94 ± 0.02^{b}	87.59 ± 0.45^{b}
T_2	0.8133 ± 0.007^{a}	0.1368 ± 0.0011^{b}	0.0267 ± 0.0002^{b}	1.81 ± 0.01^{b}	88.59 ± 0.49^{b}
T_3	0.8233 ± 0.003^{a}	0.1384 ± 0.0015^{a}	0.0270 ± 0.0001^{a}	1.54 ± 0.02^{a}	90.08 ± 0.93^{a}
T_4	0.7467 ± 0.023^{c}	0.1190 ± 0.0030^{c}	0.0242 ± 0.0008^{c}	2.21 ± 0.03^{c}	77.03 ± 2.52^{c}
p-value	0.021	0.000	0.011	0.000	0.001

Note: *Mean values with different superscripts in a column differs significantly at p<0.05 (n=3)*

4.6.4 Histological analysis of selected fish species

The histological structure of fish larvae (35 DPH) has been explained in Plate 52 and 53 for *D. dangila* and in *P. sophore*, respectively. Histological analyses showed significant difference (*p<0.05*) in number of villi, villi length formed across anterior and posterior margin of intestine and muscular thickness in both *D. dangila* and *P. sophore* (Table 35). The average number of villi and villi length was found to be increased in both *D. dangila* (34.11±4.21 and 214.41±14.25 μm, respectively) and *P.*

sophore (36.52±21.12 and 198.52±15.56 μm, respectively) significantly than control group where villi number and villi length were 29.20±6.14 and 178.24±21.45 μm in *D. dangila*, respectively (Plate 54 A and B). In case of *P. sophore*, villi number and villi length were 26.41±23.43 and 167.34±23.12 μm respectively (Plate 55 A and B). The muscular thickness was reduced significantly in *D. dangila* intestine (32.41±2.71 μm) when compared with control (34.11±2.14 μm) but significant difference in muscular thickness was not observed in *P. sophore* in treatment (29.20±6.14 μm) and control group (30.41±1.34 μm). Further, the histological sectioning confirmed that the intestinal villi get damaged and showed abnormal structures with superfluous mucus secretion at higher dosage of *Lactobacillus sp.* in *D. dangila*. In contrast, normal structures were observed in *P. sophore* intestinal sections (Plate 56 A and B).

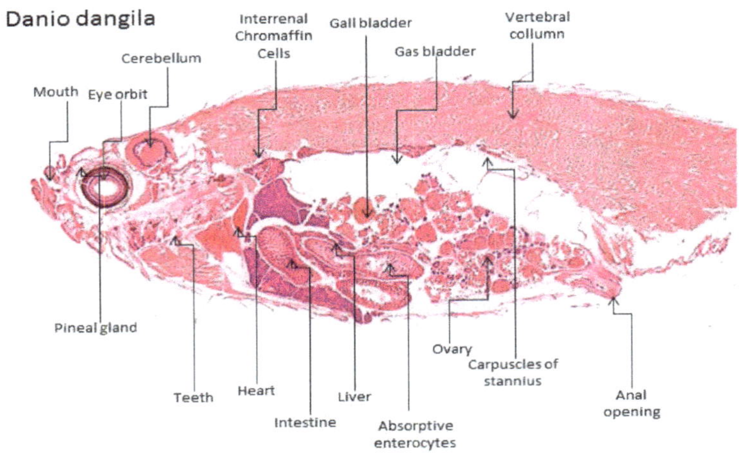

Plate 52: Histological appearance of *Danio dangila*

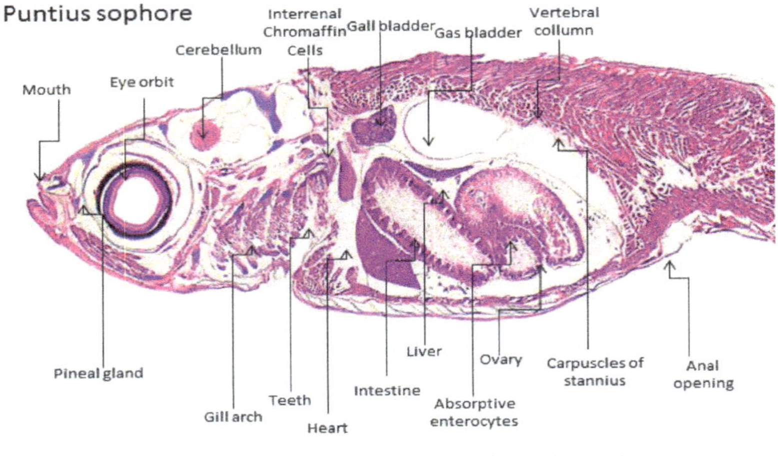

Plate 53: Histological appearance of *Puntius sophore*

Table 35: Histological analysis of intestine of *D. dangila* **and** *P. sophore* **larvae**

		Control	After 31 days	p-value
Villi Length (µm)	*D. dangila*	178.24±21.45^{b}	214.41±14.25^{a}	0.024
	P. sophore	167.34±23.12^{b}	198.52±15.56^{a}	0.038
Muscular Thickness (µm)	*D. dangila*	34.11±2.14^{b}	32.41±2.71^{a}	0.01
	P. sophore	30.41±1.34^{a}	29.62±0.98^{a}	0.047
Average Number of Villi (Anterior and Posterior)	*D. dangila*	29.20±6.14^{b}	34.11±4.21^{a}	0.034
	P. sophore	26.41±23.43^{b}	36.52±21.12^{a}	0.034

Note: *Mean values with different superscripts in a row differs significantly at* $p<0.05$

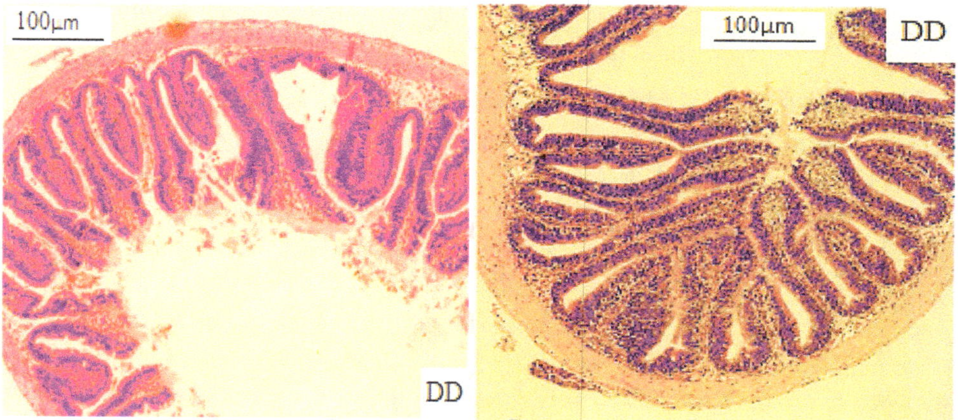

Plate 54: Histological examination of intestinal villi in *D. dangila*.
A. Control; B. Treatment

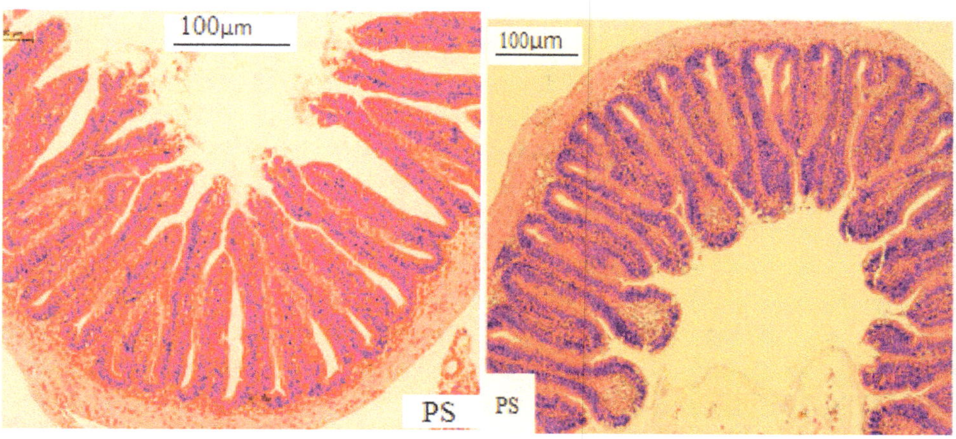

Plate 55: Histological examination of intestinal villi in *P. sophore*.
A. Control; B. Treatment

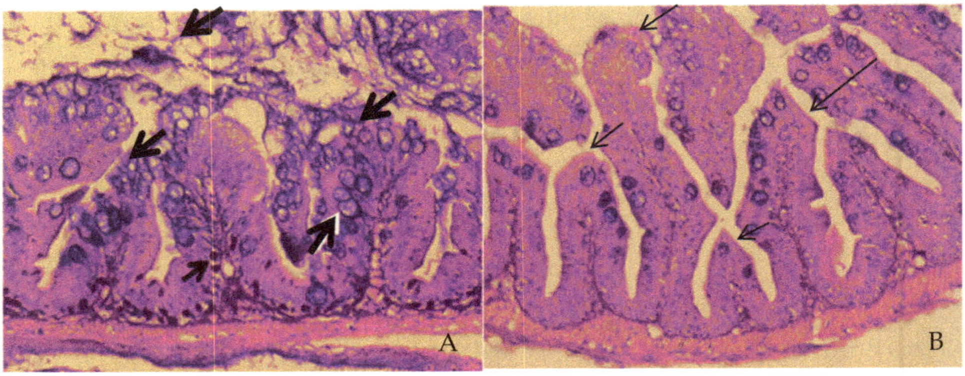

Plate 56A: Abnormal villi **Plate 56B:** Normal villi

(Harri's Haematoxylin and Eosin counterstain (HE) staining, Scale 100 μm at 40x magnification)

4.7 Larval rearing of *Danio dangila* and *Puntius sophore*

4.7.1 Effect of rearing system and stocking density on *Danio dangila*

The effect of rearing system and stocking density through growth performance indices for *D. dangila* has been described (Table 36 and Fig. 22 to 25). The fish reared in nursery raceway system exhibited significantly higher ($p<0.05$) survival rate (62.86±0.123%) compared to FRP tanks based rearing system (60.53±0.123%). In the case of stocking density, significantly higher survival rate was observed in 30 nos. L^{-1} stocking density (81.00±0.783%) followed by 10 nos. L^{-1} (80.55±0.783%). Whereas, significantly lower survival rate was observed in 70 nos. L^{-1} stocking density (28.07±0.783%) (Fig. 22). The results of two-way ANOVA revealed that, rearing system and stocking density had significant interaction effect (p = 0.041). The final weight was significantly ($p<0.05$) higher in fishes reared in nursery raceways (1.229 ±0.031 g) than FRP tanks (1.078±0.031 g). By showing significant interaction effect of rearing system and stocking density (p = 0.043), the higher mean weight was observed in fishes stocked at 30 nos. L^{-1} (1.323±0.044 g) followed by 10 nos. L^{-1} (1.272±0.044 g). While, the lower final weight was observed at 70 nos. L^{-1} stocking density (0.945±0.044 g) (Fig. 23) (Plate 57). The results of two-way ANOVA showed significant interaction effect of rearing system and stocking density (p = 0.038) exhibiting significantly higher ($p<0.05$) average daily gain (0.039±0.001 g) in nursery raceway based system followed by FRP tanks (0.034±0.001 g). In case of stocking density, highest average daily gain was observed at 10 nos. L^{-1} and 30 nos. L^{-1} (0.040±0.001 g and 0.042±0.001 g) with no significant difference followed by 50 nos. L^{-1} (0.034±0.001 g). While the lowest average daily gain was observed at 70 nos. L^{-1} stocking density (0.029±0.001 g) (Fig 24). In case of specific growth rate, fishes reared in nursery raceways (0.132±0.001% day^{-1}) showed significantly higher specific growth rate ($p<0.05$) followed by FRP tanks (0.128±0.001% day^{-1}). In comparison with stocking density, same but significantly higher specific growth rate was observed at 10 nos. L^{-1} and 30 nos. L^{-1} stocking densities (0.134±0.001% day^{-1}) followed by 50 nos. L^{-1} stocking density (0.130±0.001% day^{-1}) (Fig. 25). The results of two-way ANOVA revealed that, rearing system and stocking density had significant interaction effect ($p = 0.016$) on specific growth rate of *D. dangila* (Plate 57).

4.7.2 Effect of rearing system and stocking density on *Puntius sophore*

The effect of rearing system and stocking density through growth performance indices for *P. sophore* has been described (Table 37 and Fig. 26 to 29). The fish reared in nursery raceway system exhibited significantly higher ($p<0.05$) survival rate (76.30±1.18%) compared to FRP tanks based rearing system (68.06±1.18%). In the case of stocking density, significantly higher survival rate was observed in 30 nos. L^{-1} stocking density (91.79±1.67%) followed by 10 nos. L^{-1} (91.72±1.67%). Whereas, significantly lower survival rate was observed in 70 nos. L^{-1} stocking density (43.84±1.67%) (Fig. 26). The results of two-way ANOVA revealed that, rearing system and stocking density had significant interaction effect ($p = 0.00$). The final weight was significantly ($p<0.05$) higher in fishes reared in nursery raceways (1.062±0.021 g) than FRP tanks (0.920±0.021 g) (Plate 58). By showing significant interaction effect of rearing system and stocking density ($p = 0.00$), the higher mean weight was observed in fishes stocked at 10 nos. L^{-1} (1.097±0.029 g) followed by 30 nos. L^{-1} (1.022±0.029 g). While, the lower final weight was observed at 70 nos. L^{-1} stocking density (0.883±0.029 g) (Fig. 27). The results of two-way ANOVA showed significant interaction effect of rearing system and stocking density ($p = 0.00$) exhibiting significantly higher ($p<0.05$) average daily gain (0.034±0.001 g) in nursery raceway based system followed by FRP tanks (0.029±0.001 g). In case of stocking density, highest average daily gain was observed at 10 nos. L^{-1} (0.0349±0.001 g) followed by 30 nos. L^{-1} (0.0325±0.001 g). While the lowest average daily gain was observed at 70 nos. L^{-1} stocking density (0.028±0.001 g) (Fig. 28). In case of specific growth rate, fishes reared in nursery raceways (0.138±0.001% day^{-1}) showed significantly higher specific growth rate ($p<0.05$) followed by FRP tanks (0.133±0.001% day^{-1}). In comparison with stocking density, higher specific growth rate was observed at 10 nos. L^{-1} stocking density (0.141±0.001% day^{-1}) followed by 30 nos. L^{-1} (0.140±0.001% day^{-1}). While the lowest specific growth rate was observed in 70 nos. L^{-1} stocking density (0.120±0.001% day^{-1}) (Fig. 29). The results of two-way ANOVA revealed that, rearing system and stocking density had significant interaction effect ($p = 0.003$) on specific growth rate of *P. sophore* (Plate 58).

Table 36: Growth performance of *D. dangila* larvae reared in two different systems

	Survival (%)	Final weight (g)	Average daily gain (g)	Specific growth rate (%)
Rearing system				
FRP tanks	60.53^{b}	1.078^{b}	0.034^{b}	0.128^{b}
Nursery Raceways	62.86^{a}	1.229^{a}	0.039^{a}	0.132^{a}
SEM	0.123	0.031	0.001	0.001
p-value	0.02	0.003	0.003	0.016
Stocking density (nos L^{-1})				
10	80.55^{a}	1.272^{b}	0.040^{a}	0.134^{a}

	Survival (%)	Final weight (g)	Average daily gain (g)	Specific growth rate (%)
30	81.00[a]	1.323[a]	0.042[a]	0.134[a]
50	58.05[b]	1.073[b]	0.034[b]	0.130[a]
70	28.07[c]	0.945[c]	0.029[b]	0.123[b]
SEM	0.783	0.044	0.001	0.001
p-value	0.000	0.000	0.000	0.000
Rearing system × Stocking density (nos L^{-1})				
p-value	0.041	0.043	0.038	0.022

Note: *Mean values with different superscripts in a column differs significantly at p<0.05*

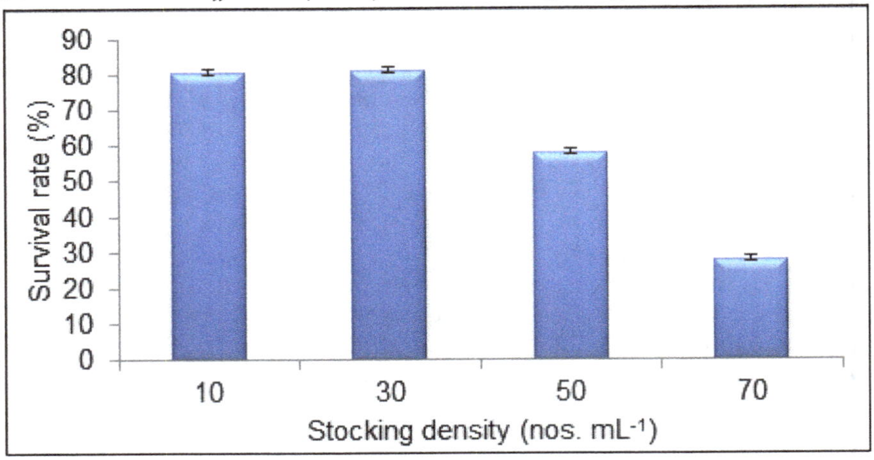

Fig. 22: Effect of stocking density on survival rate (%) of Danio dangila in nursery raceways

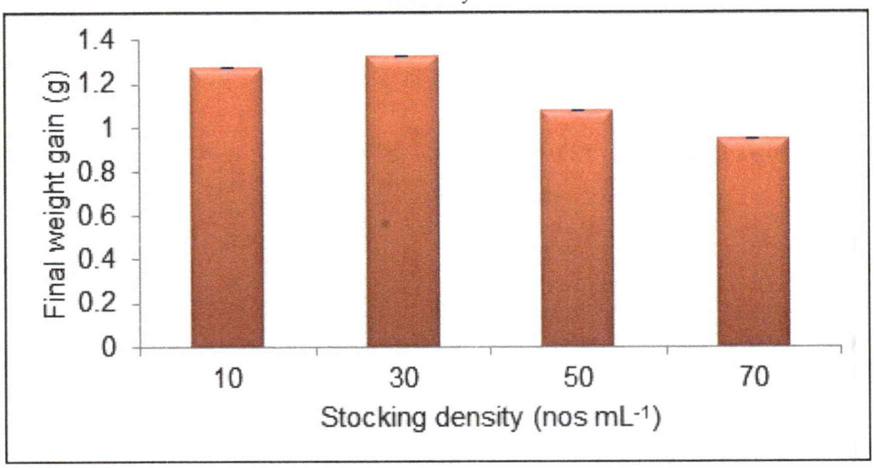

Fig. 23: Effect of stocking density on final weight gain (g) of *Danio dangila* in nursery raceways

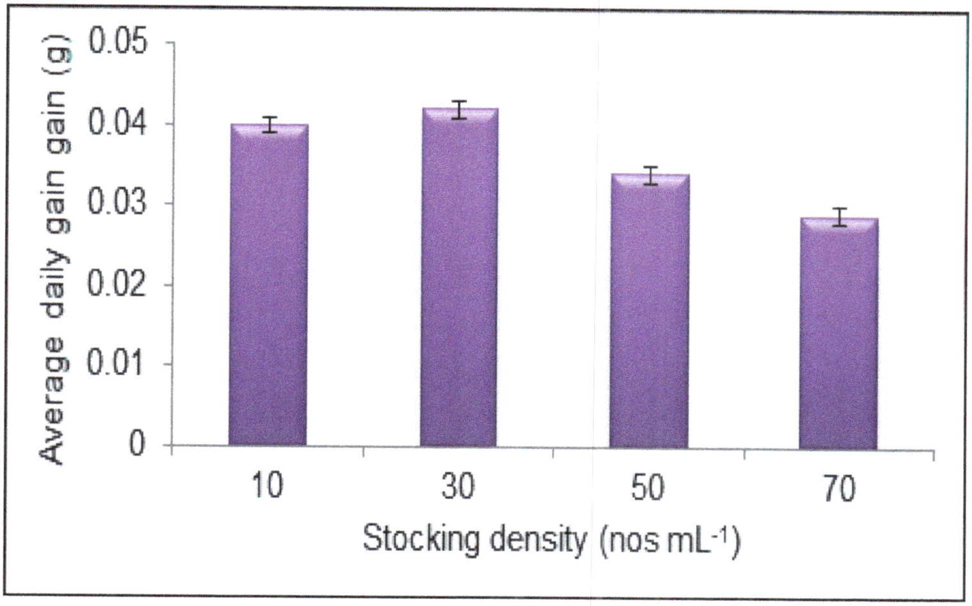

Fig. 24: Effect of stocking density on average daily gain (g) of *Danio dangila* in nursery raceways

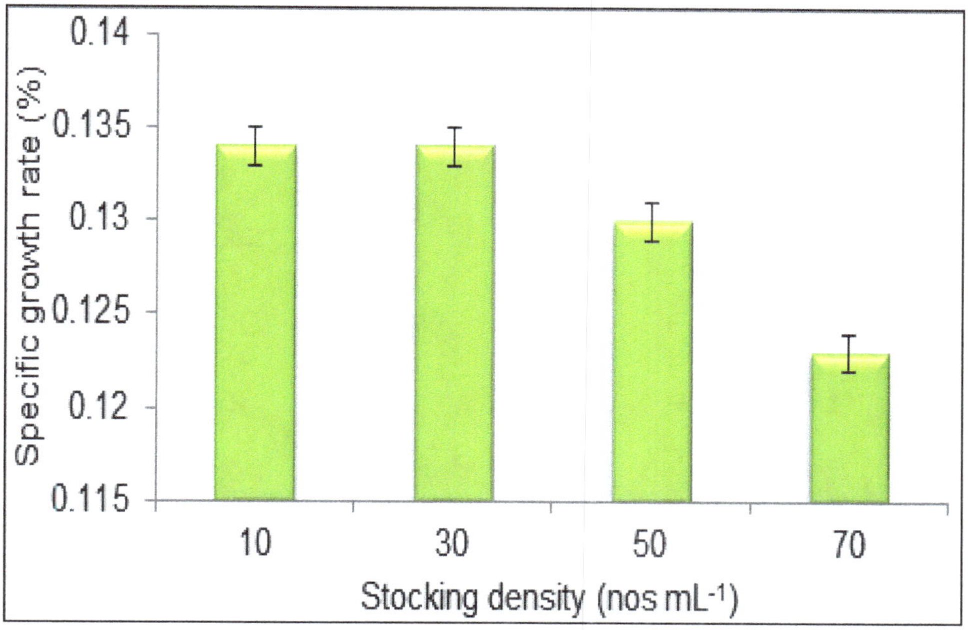

Fig. 25: Effect of stocking density on specific growth rate (%) of Danio dangila in nursery raceways

Plate 57: Larval development of *Danio dangila*

Table 37: Growth performance of *P. sophore* larvae reared in two different systems.

	Survival (%)	Final weight (g)	Average daily gain (g)	Specific growth rate (%)
Rearing system				
FRP tanks	68.06[b]	0.92[b]	0.029	0.133
Nursery Raceways	76.30[a]	1.062[a]	0.034	0.138
SEM	1.18	0.021	0.001	0.001
p-value	0.000	0.000	0.000	0.003
Stocking density (nos L^{-1})				
10	91.72[a]	1.097[a]	0.0349[a]	0.141[a]
30	91.79[a]	1.022[a]	0.0325[a]	0.140[a]
50	61.37[b]	0.955[b]	0.0304[b]	0.136[b]
70	43.84[c]	0.883[c]	0.0276[c]	0.120[c]
SEM	1.67	0.029	0.001	0.001
p-value	0.000	0.001	0.000	0.000
Rearing system × Stocking density (nos L^{-1})				
p-value	0.001	0.030	0.033	0.041

Note: *Mean values with different superscripts in a column differs significantly at $p<0.05$*

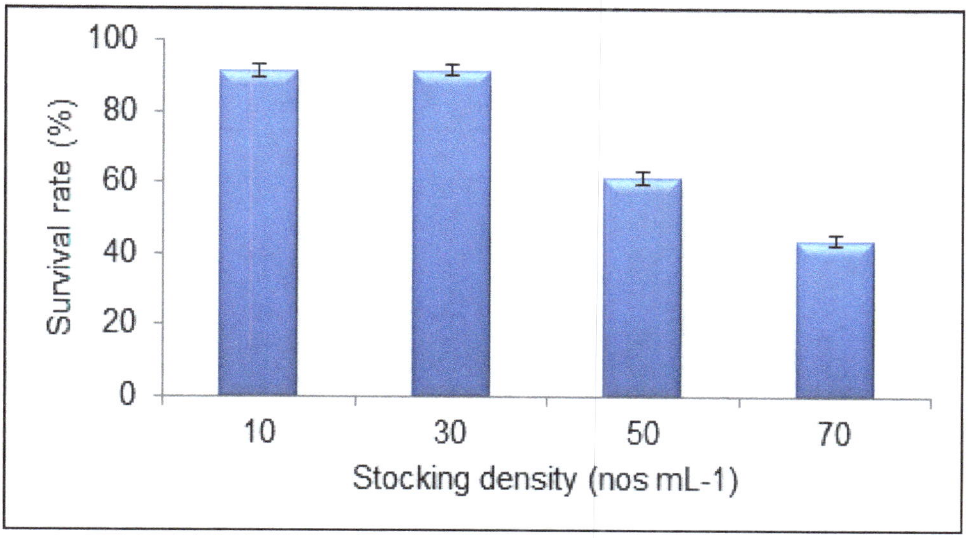

Fig. 26: Effect of stocking density on survival rate (%) of *Puntius sophore* in nursery raceways

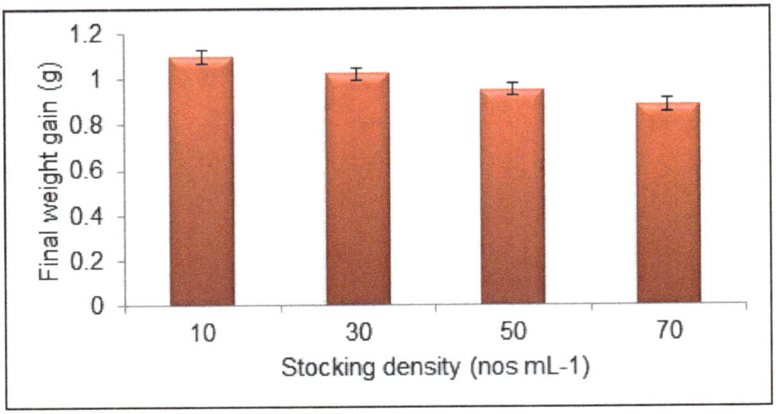

Fig. 28: Effect of stocking density on average daily gain (g) of *Puntius sophore* in nursery raceways

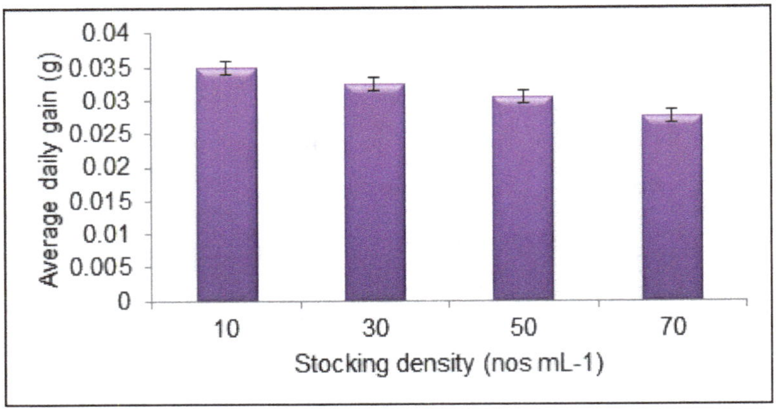

Fig. 27: Effect of stocking density on final weight gain (g) of *Puntius sophore* in nursery raceways

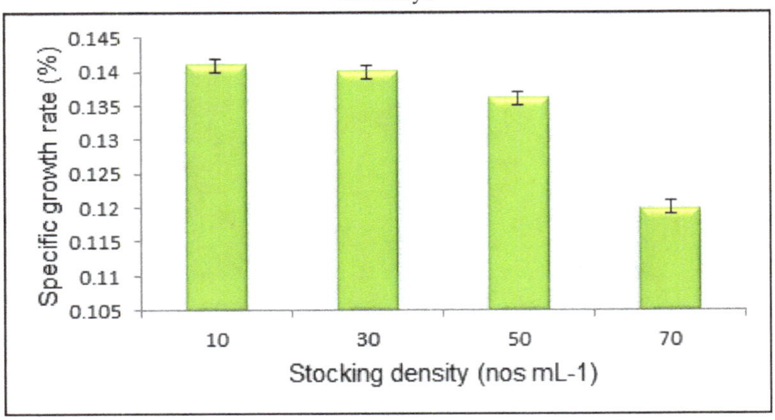

Fig. 29: Effect of stocking density on specific growth rate (%) of *Puntius sophore* in nursery raceways

Plate 58: Larval development of *Puntius sophore*

CHAPTER – 5

Discussion

In the present study, induced breeding of North-East ornamental fish species, *Danio dangila* and *Puntius sophore*, mass production of important freshwater live food species, *Brachionus calyciflorus* and formulation of probiotic incorporated encapsulated feed were performed with a view to standardize breeding technologies of indigenous ornamental fishes and larval rearing technologies in advanced culture systems like raceways with optimum stocking density and standard management practices.

5.1 Induced breeding of North-East region based ornamental fishes, Moustached Danio, *Danio dangila* and Sophore barb, *Puntius sophore* in captivity

5.1.1 Induced breeding of Moustached Danio, *Danio dangila*

The present study demonstrated successful breeding of North-East region based endemic ornamental fish, *Danio dangila* at Southern part of India in Tamil Nadu with the use of synthetic hormones along provided insight on the embryonic developmental stages. The results of the present study suggested that the molecular analogue based synthetic hormone, WOVA-FH at a dose of 0.5 mL kg^{-1} for female and 0.3 mL kg^{-1} for male were optimal to induce *D. dangila* for getting higher spawned egg output (1002.50±52.6), fertilization rate (84.0±2.72%) and hatching rate (86.5±2.9%) within optimum latency period (6.00 – 8.00 hours) as compared to other hormones and doses. In similar way, Motilan et al., (2013 and 2014) has successfully bred *Puntius manipurensis* and *Puntius chola* from North-East India using WOVA-FH at a dose of 0.4 mL kg^{-1} for both the species with an average number of spawned eggs 6218.75±32.75 and 893.34±41.34, respectively.

Mahapatra (2010) also succeeded in induced breeding and rearing of indigenous ornamental fish, *P. shalynius* using ovaprim at a dose of 0.4 mL kg^{-1}. In similar way, Udit *et al.* (2014) used ovatide at 0.2 mL 200 gm^{-1} for induced breeding of *Puntius sarana* with 78500 egg output. Chaudhari and Alikunhi (1957 and 1962) spawned Indian major carps and *Puntius sarana* through hypophysation technique while Bhuiyan et al., (2006) used different doses of pituitary gland extract for breeding the *Puntius gonionotus* which resulted in varying eggs output and further reproductive parameters like fertilization and hatching rates.

The latency period or response time is often related to the water temperature which decreases with the increase in temperature (Kucharczyk et al., 2005). Differences in the latency time during the breeding of Tench (*Tinca tinca*) were observed in the case of different spawning agents (Jamorz et al., 2008). In this study, latency period of 6-8 hours at ambient water temperature (21-27°C) was similar to the reports of Begum et al., (2009) and Singh et al., (2015). Mangesh et al., (2017) documented different latency period (6-12 hrs) in *Hemichromis bimaculatus* at different dosage of WOVA- FH. Similar observations were reported by Pandey et al., (2002) in *Labeo rohita*. Behera et al., (2010) also reported longer latency period in low dose of synthetic hormone, ovaprim and ovatide, in *Labeo bata*. According to Billiard et al., (1984) and Peter et al., (1986), differences in dose requirement may be attributed to varied level of dopamine activity in fishes. Das et al., (1994) and Chatterjee *et al.,*(2013) have documented injection of WOVA-FH at a dose of 0.35 mL kg^{-1} of body weight of a mature female Tawes (*Puntius javanicus*) and Angel fish (*Pterophyllum scalare*) gave better spawning within 4-5 hrs.

The difference found in the induced spawning of fish with different hormones and dosages were may be due to target organs of the hormones (Billiard et al., 1984). Human chorionic gonadotropin (HCG) elicits shorter latency period than Luteinizing hormone-releasing hormone (LHRH) because of the difference in mode of action. HCG forces gonad to open up even though it is not developed. Whereas, LHRH acts on the brain to have well matured eggs (Zohar and Mylonas, 2001). Preceding investigations have shown the potential actions of synthetic analogues combined with dopamine antagonists like pimozide, domperidone or metachlopramide (Billard et al., 1983 and Peter et al., 1988) which formed the basis for the use of gonadotropic signaling molecular analogues like WOVA-FH, Ovatide and ovaprim for induced breeding of fish species. Since, the activity of dopamine antagonist varies with fish species, the response to synthetic hormones with different antagonist is likely to be different in different species (Nandeesha et al., 1990).

In the present study, longer latency period might be due to the use of less quantity of hormone which may not be sufficient to induce female fish. Similarly, Kumar et al., (2018) reported longer latency period in *Mystus gulio* when induced with lower amount of hormone. Over-dosing of the inducing agents caused early milting; resulting in poor fertilization and under-dosing caused late inducement in males. Similar findings were reported by Mangesh et al., (2017) in *Hemichromis bimaculatus* and Routrey et al., (2007) in carps. Behera et al., (2007) reported that the eggs output per female, fertilization and hatching rate were highest using ovaprim at a dose of 0.5 mL kg^{-1} for female and 0.2 mL kg^{-1} for male body weight but gradually decreased with increased dosage. In case of Ovatide, fertilization and hatching rate were highest at 0.4 ml kg^{-1} of female and 0.2 ml kg^{-1} of male body weight in *L. bata*. In similar manner, Sukumaravin et al., (2007) and Motilan et al., (2014) found increased percentage of egg production, fertilization and hatching rate in Thai carp and *Pethia manipurensis* using WOVA-FH.

Optimal range of water quality parameters recorded in the breeding tanks also might have contributed to increased hatching rates. Behera et al., (2007), Rath et al., (2007) and Pranob Das et al., (2016) reported that the hatching rate of *L. bata,* IMC and *O. belangeri* was highest when induced with synthetic hormones along with optimum range of water quality in the experimental ponds. Rath *et al.* (2007) also reported higher hatching rate in Indian Major Craps (IMC) administered with

WOVA-FH at 0.4-0.5 mL kg^{-1} for female when compared with other hormones with optimum water quality parameters.

5.1.2 Induced breeding of Sophore barb, *Puntius sophore*

The present study demonstrated successful breeding of North-East region based indigenous ornamental fish, *Puntius sophore*, at Tamil Nadu in captivity with the use of different hormones along with its embryonic developments. The results of the present study revealed that the WOVA-FH at a dose of 0.4 mL kg^{-1} for female and 0.3 mL kg^{-1} for male are sufficient to induce *P. sophore* for getting higher egg output (881.50±85.8), fertilization (78.50±2.72%) and hatching rate (83.75±2.39%) as compared to other hormones and doses. In similar way, Motilan *et al.* (2013 and 2014) had successfully bred *Puntius manipurensis* and *Puntius chola* using WOVA-FH at a dose of 0.4 mL kg^{-1} (6218.75 ± 32.75). Dey *et al.* (2016) also reported that the higher egg output (5020 eggs $female^{-1}$) in *Barilius barila* when induced with WOVA-FH at a dose of 0.5 mL kg^{-1}. Singh et al., (2015) and Chaturvedi et al., (2016) reported higher success rate in Climbing perch (*Anabus testudeneus*), when induced with WOVA-FH at a dose of 0.4 mL kg^{-1} when compared to other hormone (ovaprim). Mangesh et al., (2017) documented difference in eggs output at different dosage of WOVA- FH in African jewel fish, *Hemichromis bimaculatus*.

Mahapatra (2010) suggested a dose of 0.4 mL kg^{-1} of Ovaprim for induced breeding of indigenous ornamental fish, *P. shalynius* while Udit *et al.* (2014) documented a dose of 1.0 mL kg^{-1} of ovatide for induced breeding of *Puntius sarana* generated higher success rate. Over mLHRH-A (ovapel), the sLHRH-A was more potent as it is a composite mixture of sGnRH-A with dopamine antagonists dissolved in non-toxic calibrated amount of organic solvent. The increased effectiveness of sGnRH-A is attributed to its higher affinity for the binding site of pituitary (Peter et al., 1986). Kahkesh et al., (2010) reported highest spawning success (87.5%) in *Barbus sharpeyi* when induced with sLHRH-A with crude pituitary extract. Chaudhari and Alikunhi (1957 and 1962) and Bhuiyan et al., (2006) were successfully bred Indian major carps, *Puntius sarana* and *Puntius gonionotus*, respectively, through hypophysation technique with different doses of pituitary gland extract.

The latency period or response time is often related to the water temperature, synthetic hormone activity and the dose applied. The latency period decreases with an increase in temperature while improper dose of synthetic hormones leads to shorter or longer latency period (Kucharczyk et al., 2005). In this study, latency period of 6-8 hours at ambient water temperature (21-27°C) was similar to the reports of Pandey et al., (2002), Begum *et al.* (2009), Dey *et al.* (2016) and Singh *et al.* (2015) in barbs. Differences in the latency time during the breeding of Tench (*Tinca tinca*), Benni (*Barbus sharpeyi*) were observed in the case of different spawning agents (Jamorz et al., 2008; Kahkesh et al., 2010).

The latency period in *P. sophore* was prolonged, considerably, when induced with higher doses of synthetic hormone. Behera et al., (2010) also reported longer latency period in higher dose of synthetic hormone, ovaprim and ovatide, in *Labeo bata*. According to Billiard et al., (1984) and Peter et al., (1986) stated differences in dose requirement may be attributed to varied level of dopamine activity in fishes. Das et al., (1994) and Chatterjee et al., (2013) documented longer latency period (6.45 Hours) in Tawes (*Puntius javanicus*) and Angel fish (*Pterophyllum scalare*) when induced with WOVA-FH at a dose of 0.6 mL kg^{-1}.

Since variations in the environmental condition of the fish exerts certain stresses, which may have adverse effects on gonadal maturity. Hence, induced breeding of fish through hormonal treatment appears to be relevant in conservation through mass production of the species (Silva et al., 2002 and Quintana et al., 2004). The difference found in the induced spawning of fish with different hormones could be due to target organs. HCG elicits shorter latency period than LHRH because of the difference in mode of action by forcing gonad to open up even though under developing stage. However, LHRH acts on the brain to ooze out only well matured eggs (Zohar & Mylonas, 2001).

In the experiment, the administered hormone and its dose apparently affected the rate of fertilization and hatching rate. Highest rate of fertilization (78.50±2.72%) and hatching (83.75±2.39%) in *P. sophore* was observed with the administration of WOVA-FH at 0.4 mL kg^{-1} for female and 0.2 mL kg^{-1} for male. Over-dosing of the inducing agents caused early milting resulting in poor fertilization and under-dosing caused late inducement in males. Similar findings were reported by Routrey et al., (2007) in barbs and Mangesh et al., (2017) in African cichlid, *Hemichromis bimaculatus*. Behera et al., (2007) reported that the egg output per female, fertilization and hatching rate was highest using ovaprim at a dose of 0.5 mL kg^{-1} of female and 0.2 mL kg^{-1} of male body weight while in case of ovatide it was of 0.4 mL kg^{-1} of female and 0.2 mL kg^{-1} of male body weight in *L. bata* but considerably reduced when induced with higher doses. In similar manner, Sukumaravin et al., (2007) and Motilan et al., (2014) found increased percentage of egg production, fertilization and hatching rate in Thai carp, *Puntius gonionotus* and *Puntius manipurensis* using WOVA-FH when induced at 0.4 mL kg^{-1}.

The dose of hormone obviously affected fertilization and hatching rate in *P. sophore* when induced with different hormones at various doses. The differences found in the induced spawning of fish could be attributed to target organs of the hormones in which Human chorionic gonadotropin (HCG) elicits shorter latency period as it forces gonad to open up even though it is not developed (Billiard et al., 1984). On the other hand, Luteinizing hormone-releasing hormone (LHRH) acts on the brain to have well matured eggs (Zohar and Mylonas, 2001). Preceding investigations suggested that the potential actions of synthetic hormones with dopamine antagonists like pimozide, domperidone or metachlopramide for induced breeding of fish species (Billard et al., 1983 and Peter et al., 1988). Since, the activity of dopamine antagonist varies with fish species, the response to synthetic hormones with different antagonist is likely to be altered in different species (Nandeesha et al., 1990).

Optimal range of water quality parameters recorded in the breeding tank also might have contributed to increased hatching rates. Bhuiyan et al., (2006), Behera et al., (2007), Rath et al., (2007) and Pranob Das et al., (2016) reported, highest hatching rate in *Labeo bata,* IMC, *Osteobrama belangeri* and *Puntius gonionotus,* respectively, when induced with synthetic hormones along with optimum range of water quality in the experimental ponds. Rath *et al.* (2007) also reported higher hatching rate in IMC administered with WOVA-FH at a dose of 0.4-0.5 mL kg^{-1} of female body weight with optimum water quality parameters and same was supported by Singh et al., (2015) and Bhattacharya and Homechaudhury, (2009) for *Anabus testudineus*.

Although the majority of freshwater fish involved in the aquarium fish trade are from captive-bred sources, significant numbers are still removed from the wild (Rahdari et al., 2014). Therefore, captive breeding can play a great role in increasing the population of these species which can be used by the aquarium keepers to avoid overexploitation of the species from wild for its conservation along with its habitat.

5.2 Mass production of freshwater Rotifer, *Brachionus calyciflorus*

In the present study, mass production of freshwater rotifer, *Brachionus calyciflorus* was done with a view to find out effective food type preferred by rotifers, intensity of feeding for rotifer production and biomass production and pigment enhancement of selected food type, *Chlorella vulgaris,* using internal photobioreactor. Further, the study marched for mass production of freshwater rotifer, *Brachionus calyciflorus,* in advanced aquaculture systems like Recirculatory Aquaculture System (RAS) with a view to find out effective food type and to standardize daily algal cell, *Chlorella vulgaris,* requirement of individual rotifer with standard management practices by optimizing effective harvesting rate for maintaining rotifer culture stable throughout the season with zero water exchange.

5.2.1 Mass production of freshwater Rotifer, *Brachionus calyciflorus* in outdoor culture system

5.2.1.1 Effect of different food types on Rotifer, *Brachionus calyciflorus* production

Different food types showed significant difference ($p<0.05$) in rotifer production in outdoor culture system. Higher growth was observed in tanks fed with *C. vulgaris* (180.33±2.18 nos. mL^{-1}) on 8th day. Similarly, Ashraf et al., (2010) reported that *C. vulgaris* showed the maximum rotifer production followed by baker's yeast and *Spirulina* sp. while Kennari et al., (2008) observed that *B. calyciflorus* showed better production and fatty acid content when fed with *Chlorella* sp. compared to *Scenedesmus* sp. This indicates that size, shape and nutrient composition of *Chlorella vulgaris* made it more accessible for the rotifer due to their mouth diameter. In contrast, Ajah (2010) observed that giant adult rotifers (*B. quadridentatus*) gave greater preference to *Scenedesmus quadricauda* and *Eudorina elegans* than *C. vulgaris* which may be due to their larger size hence, prefer larger sized food type.

Almost equal selection of *C. vulgaris* and Baker's yeast diets led to uniform productivity, irrespective of alga and food type showing effective production of rotifers (132.00±9.03 nos. mL^{-1}) compared with other species (*S. major* and *S. obliquus*). Perhaps, *B. calyciflorus* preferred Baker's yeast possibly due to problems related with long, thread like structure and uneven size of algae which made the feeding difficult (Odo et al., 2015). Even though, Baker's yeast was a good food for rotifer, its lower performance compared with *C. vulgaris* could be attributed due to vitamin deficiency particularly, vitamin B_{12} and, comparatively costlier than *Chlorella* culture (Zou and Davis, 2000). Sarma *et al.* (2001) reported that *B. calyciflorus* grew better on *Chlorella sp.* alone while *B. patulus* grew equally well when fed with *Chlorella sp.* or mixed with *Saccharomyces cerevisiae* in equal proportions. Therefore, it has been opined that the *C. vulgaris* acted as an excellent food for rotifer production compared to other species and Baker's yeast, because of their small size, nutrient contents and higher multiplication rate makes it more accessible to prey.

5.2.1.2 Effect of feeding intensity on rotifer production

Many studies suggested that the rotifer growth is limited in natural water bodies due to less food availability and choice of food type (Sarma et al., 2011). The food type efficiently preferred by rotifer was further used to find out the extent of concentration needed to maximize rotifer production. Feeding interval at 2× and 3× day^{-1} showed significant effect on rotifers production ($p<0.05$). Rotifers fed at the intensity of 3× per day with *C. vulgaris* showed highest production (256.33±3.8 nos. mL^{-1}) on 8th day than the tanks fed at 2× per day (186.67±1.6 nos. mL^{-1}). In support with the present findings, Oie *et al.* (1994) stated that quality and quantity of food is the most important criteria effecting rotifers production. Odo et al., (2015) found higher rotifer production (213 nos. mL^{-1}) when fed with *Chlorella* 3× day^{-1} compared with 2× day^{-1} (166 nos. mL^{-1}) on 8th day onwards in batch culture system while Chew and Lim (2006) reported decrease in rotifers production (41 nos. mL^{-1}) with decreased daily feeding. Similarly, Ajah (2008) observed continuous increase in the population of *Asplanchna priodonta* upon increase in algal population.

However, in contrast, Lavens and Sorgeloos (1996) found that decrease in cell density did not decrease the rotifer number up to 3rd day after inoculation. By nullifying Lavens and Sorgeloos (1996) statement, Ozdemir and Ciltas, (2010) reported that the food limitation caused pressure on reproduction and increased mortality in outdoor culture of rotifers. They also reported highest rotifer density (269±40.12 nos. mL^{-1}) in a group fed with *Chlorella sp.* at 12 hours intervals and when the length of feeding intervals were increased upto 24 hours, the rotifer density started decreasing in outdoor culture tanks. Similar findings were reported by Suantika *at al.,* (2000), Arimoro, (2006) and Ashraf et al., (2010).

5.2.2 Biomass production and pigment enhancement of single cell algae, *Chlorella vulgaris* in internal photobioreactor

5.2.2.1 Effect of light wavelength on biomass production of *Chlorella vulgaris*

In the present study, overall effect of different light wavelengths applied at 5 kLux showed significance difference in biomass production of *Chlorella vulgaris* ($p<0.5$). The highest cell count (65.220±0.17×10^6 cells mL^{-1}) and specific growth rate (0.79±0.32 day^{-1}) of *C. vulgaris* was observed in green light. Similar reports were given by Ankita et al., (2013) and concluded that energy content of green light and blue light is greater that red light probably causing growth increament. A research using green light under different light intensities (3, 5 and 7 kLux) and lighting periods (8:16, 12:12 and 16:8) (light:dark) showed that a long time lighting of green wavelength at 5 kLux showed significant increase in *C. vulgaris* biomass (2.025 g L^{-1}) (Zehnder and Gorham, 1960; Mayo, 1997). Rosenberg et al., (2008) reported higher specific growth rate of *Chlamydomonas reinhardtii* in green light.

On the other hand, the lowest cell count ($43.692 \pm 0.13 \times 10^6$ cells mL^{-1}) and growth rate (0.22 ± 0.32 day $^{-1}$) of *C. vulgaris* was observed in red light compared with other sets at the end of experiment. Asuthkar et al., (2017) also reported that growth of alga (*Chlorella pyrenoidosa*) in red light was comparatively lesser than white, blue and green lights. In contrast, Wong et al., (2016) reported significant higher growth of *C. vulgaris* in cool white light cultured in BG-11 culture medium but they have not used green light.

The green light was reported to be most effective in production of cellular carbohydrate (i.e., CH_2O) derived from CO_2 (Ryu et al., 2009). Also, green photon has 20% more energy than one red photon (i.e., 680 nm) and 15.5% less energy than a blue photon (i.e., 470 nm). Therefore, higher blue light exposure may have some inhibitory effects due to higher energy content (Blair et al., 2014 and Khoeyi et al., 2012).

5.2.2.2 Effect of light wavelength on cell size, protein and pigment enhancement

Algae are not only the most auspicious feed for small organisms and shrimp/fish larvae (Felix et al., 2004) but also as a stock for biofuel production. Algae are primitive photosynthetic organism contain chlorophylls and other pigments for photosynthesis, which trap light energy and convert into chemical energy. It stores energy in the form of starch and carbohydrate (complex sugars) (Singh and Singh, 2015). They are indispensable as they produce about half the oxygen in Earth's atmosphere. Almost all algae have the ability to produce energy rich oils and several microalgae species naturally accumulate high levels of oils (Sudhakar and Premalatha, 2012). Moreover, algae are found in diverse habitats, multiply quickly and use carbon dioxide (Devaraja et al., 2017).

Several parameters affect the algal growth, reproduction, as well as lipid accumulation in algae, but role of light is very important (Wu, 2016). The growth of all algal species was found to be highly dependent on solar radiation in the natural water bodies. The number of investigators has reported the influence of different quality and quantity of light in addition to other physico-chemical parameters (Litchman and Klausmeier, 2001 and Blair et al., 2014). Daliry *et al.* (2017) examined the light fluctuation affected phytoplankton community structure and diversity. Evidence suggests that the light acts as a helping factor to cell proliferation and helps cellular respiration and photosynthesis. The growth of microalgae in light limited conditions, the cellular mechanisms progresses to produce amino acids and other essential compounds from carbon sources for cell. In contrast, the saturated illumination leads to maximum cell multiplication with increased production of sugar and starch. However, some findings suggest, a non-continuous illumination strategy, as single-celled photosynthetic culture usually occurs under dark conditions, to maintain high growth rates and reduce production costs (Al-Qasmi et al., 2012 and Hultberg et al., 2014).

Chlorella species are single cell green algae, contains photosynthetic pigment chlorophyll a and chlorophyll b. In the presence of water, sunlight, CO_2 and nutrients, chlorella cells replicates faster. Temperature and light required for the autotrophic growth of *Chlorella minutissima* was observed by Lotfi et al., (2011) by providing a 400-W phyto-claude halogen lamp in temperatures between 10°C and 35°C showed minimum irradiance to sustain net growth. Maximum specific growth rate was increased from 0.12 d^{-1} at 10°C, to 0.66 d^{-1} at 30°C. Another investigator observed optimum daytime temperature was 30°C for *Chlorella pyrenoidosa* for maximum biomass and lipid production (Lotfi et al., 2011). Response of micro algal growth to various light intensities (i.e. 400, 800, 1200, 1600, 2000, and 2400 mmol m^{-2} s^{-1}) was observed under the wavelength of red light and suggested that light intensity at 400 mmol m^{-2} s^{-1} was too low to maintain the growth of microalgae, *Chlorella vulgaris*, whereas 2400 mmol m^{-2} s^{-1} light intensity was too high to avoid photo inhibition (Cheng et al., 2013). Temperature, irradiance and light are important factors for better algal productivity. Maintenance of temperature for mass culturing is difficult. The optimum water temperature needed for cultivation of microalgae is around 25±2.0°C beyond this, micro algal cell damage or death may occur (Yun and Park, 2001). *C. vulgaris* cultured in flasks of 0.25 dm^3 with 0.1 dm^3 of nutrient media at 27°C temperature on a rotating shaker (150 rpm) using continuous illumination (200±50 mmol m^{-2} s^{-1}) showed corresponding doubling time at 33 hours (Yun and Park, 2001). *Chlorella fusca* and *Chlorella saccharophila* were grown under high (20 W m^{-2}) and low (5 W m^{-2}) intensities of white light showed no significant change in the pigment composition or distribution of pigment protein complexes in both the species (Mathias and Horst, 1990).

Light is a major factor in the process of photosynthesis to convert carbon dioxide into organic compounds, such as carbohydrates and proteins, in which water and oxygen get released (Hultberg et al., 2014). Photosynthesis and protein synthesis in microalgae is controlled by light to sustain cell growth (Carvalho et al., 2011). In the present experiment, highest protein content was observed in the algae culture with red light (0.132±0.14 mg mL^{-1}) followed by white light (0.049±0.01 mg mL^{-1}). Similar reports were postulated by Kendirlioglu and Cetin (2017) showing the highest protein content (266.04 μg mL^{-1}) in *C. vulgaris* cultured in red illumination trailed by white illumination (182.34 μg mL^{-1}). Mohsenpour et al., (2013) stated that the light is one of the most important factors controlling the increase in protein which occurred highest under red light in the first five days in *Chlorella vulgaris*.

Today, more than 40,000 microalgae have been identified by molecular and microscopic methods and a significant portion of them have been found to have high lipid content (Kendirlioglu and Cetin, 2017). Mostly, temperature effect on lipid concentration has been studied (Singh and Singh, 2015). In the present experiment, the highest lipid content was observed in algae grown under white light (0.252±0.24 mg mL^{-1}) followed by green light (0.210±0.31 mg mL^{-1}). Light intensity promotes cell growth, carbohydrate, lipid productivity, and CO_2 fixation efficiency in *Scenedismus obliquus* (Shih-Hshin et al., 2012). Yoshioka Miwa et al., (2012) reported that highest lipid content was obtained in the culture under blue intermittent light (155 mg L^{-1}) followed by constant white light (98 mg L^{-1}). Wong et al., (2016) also reported maximum lipid yield with the blue color at 34.06% with an 18.18% increase during the cultivation period. However, both the researchers were not compared lipid yield in green light illumination.

Chlorophyll (a and b) content also varied significantly with the different light wavelengths. Highest Chlorophyll a (0.177±0.17 mg mL^{-1}) and chlorophyll b (0.141±0.13mg mL^{-1}) content was observed in sets illuminated with green light followed by blue illumination. Similarly, Kendirlioglu and Cetin, (2017) reported maximum amount of chlorophyll a (0.241 mg mL^{-1}) and chlorophyll b (0.187 mg mL^{-1}) was measured in cultures illuminated with green light. As per Khalili et al., (2015), the absorption band of the chlorophyll is prominent in blue and red region of *C. vulgaris*. However, Wong et al., (2016) reported high chlorophyll content in *C. vulgaris* under blue and green illuminations in internal LEDs-illuminated photobioreactor.

Microalgae have a wide potential for valuable pharmaceuticals and pigments. One of the most important pigment is astaxanthin (3,3-dihydroxy-β,β-carotene-4,4'-dione), a red colored ketocarotenoid pigment, which is in growing demand in food, nutrition and cosmetics as it plays important role as antioxidant (Mohsenpour and Willoughby, 2013). Light is considered to be one of the most important factors playing an important role in the accumulation of astaxanthin in microalgae (Katsuda et al., 2004). In the present study, highest astaxanthine production was observed in culture illuminated with red light (0.0126±0.08 mg mL^{-1}) followed by green light (0.0104±0.02 mg mL^{-1}). Katsuda Tomohisa et al., (2006) suggest that with the use of intermittent flashing light from blue light emitting diodes has significant effect on the cell growth and astaxanthin production in *Haematococcus pluvialis*. Similarly, Kendirlioglu and Cetin, (2017) reported maximum amount of astaxanthin under blue light illumination (35.65 mg L^{-1}) followed by green (20.30 mg L^{-1}).

In addition, during cell count it was observed that the cell size of *C. vulgaris* varied significantly. In the present study, superior cell size of alga was observed with red light (5.2-9.4 μ) followed by white light (5.0-8.9 μ). Similarly, Anne et al., (2013) evaluated the effect of three different light intensities (50, 94 and 136 mmol m^{-2} s^{-1}), on the cell density and cell size of *N. oleoabundans* cultivated in a modified Bold's Basal medium and reported that the cell size varied significantly by light intensity, being higher at the lowest (10.927±1.26 μm) and medium light intensities (11.887±1.12 μm), compared to the size observed at the highest level of light intensity (5.257±1.26 μm).

5.2.2.3 Effect of common purified salt on the control of Ciliates in *Chlorella vulgaris* culture tanks

Ciliates posed serious problem in algal culture tanks throughout the culture period causing frequent crashes in algal culture tanks. The major ciliates observed in algal culture were *Lacrymaria* sp., *Paramecium* sp., *Coleps* sp. and *Dileptus* sp., which were not desired, as they compete with rotifer for feed (Hino, 1993 and Watanabe et al., 1983). Application of chemicals effectively controls ciliate contaminations within shorter time period but the residues remain as it is in the animals, may affect fish larvae (Dhert, 1996; Nhu, 2004; Arimoro, 2016 and Ashraf et al., 2010).

Common purified salt played important role in the control of Ciliates from algal culture tanks. For the doses of 20 and 40 g salt L^{-1}, no effects were observed even after 24 hours but salt at a dose of 60 g L^{-1} showed significant reduction in ciliates without affecting algal cells after 24 hours. In contrast, when salt was applied at a dose of 80 g L^{-1}, both algal cells and ciliates were found dead after 15 hours. Similarly, Menezes-Benavente *et al.* (2004) tested different concentrations of sodium chloride and reported significant reduction in ciliate quantity within one day without affecting C. *vulgaris* growth when applied at 85 g L^{-1} but decreased in C. *humicola* which indicates, different species of algae exhibit variable response to high salinity.

However, Dhert (1996) has different view point and stated that presence of ciliates in the culture medium is not necessarily harmful except at high concentrations but, he did not differentiate high and low levels of ciliates. Nhu (2004) nullifies Dhert (1996) and reported that the excessive presence of ciliates in large cultures lead to considerable reduction in yield because of higher competition of algae as a food and thereby reduces the availability of food to rotifers (Arimoro, 2006).

5.2.3 Mass production of freshwater Rotifer, *Brachionus calyciflorus* in indoor culture system

5.2.3.1 Effect of different food types on Rotifer, *Brachionus calyciflorus* production

Various mechanisms have been suggested to explain the poor survival of freshwater rotifers. Some of them are a) failure of osmoregulation, b) energy loss, c) reduced swimming rate, d) scarcity of favorite algal species, e) low hatching success and f) biochemical changes. Identification of appropriate food species and control of bio-chemical factors hinder the mass production of rotifers (Epp and Lewis 1984, Yamasaki and Hirata 1985, Frolov *et al.* 1991, Oeie and Olsen 1993).

In the present experiment, rotifer, *B. calyciflorus,* showed significant difference ($p<0.05$) in the growth rate with different food types. Algal species, *C. vulgaris* showed the maximum rotifer production (365.33±2.18 nos. mL^{-1}). Findings obtained was supported by reports given by Ashraf et al., (2010), who stated that the tanks fed with *C. vulgaris* had the maximum rotifer production (380 nos. mL^{-1}) followed by baker's yeast and *Spirulina* sp. In similar way, Kennari et al., (2008) testified that the *B. calyciflorus* had better production and fatty acid content when fed with *Chlorella* sp. compared to *Scenedesmus* sp. Odo et al., (2015) also reported that rotifers fed with *Chlorella sp.* (650 mg mL^{-1}) showed highest production (213.81±9.94 indi. mL^{-1}) rather than *Spirulina sp.* Rotifers preferred primarily *C. vulgaris* and then Baker's (245.33±5.36 nos. mL^{-1}) yeast rather than *S. major* (86.00±0.21 nos. mL^{-1}) and *S. obliquus* (22.67±10.80 nos. mL^{-1}). This might be due to long, spiral, thread like structures and irregular shapes of alga, make difficulties to ingest them (Sarma et al., 2001 and Ajah, 2010).

Though rotifers preferred Baker's yeast, its lower performance compared with *C. vulgaris* could be attributed to vitamin deficiency, B_{12}, (Treece and Davis, 2000) and change in water quality. This indicated that size, shape and nutrient composition of *C. vulgaris* made it more accessible to prey because it suits to mouth diameter of rotifers (Abou-Shanab et al., 2016 and Lokaj *et al.*,2016) and also water quality can be maintained by avoiding yeast based products (Suantika et al., 2003).

5.2.3.2 Estimation of algal, *Chlorella vulgaris*, consumption rate by individual rotifer

In the present experiment, preferred food type, *C. vulgaris* consumption rate by individual rotifer was estimated and significant difference was observed in rotifer production. The study revealed that almost equal rotifers recovery from two tanks (544.63±44.03 indi. mL^{-1}) and (541.33±21.18 indi. mL^{-1}) when applied with algal cells at the density of $121.05±0.95×10^3$ cell mL^{-1} and $79.60±1.27×10^3$ cells mL^{-1} with the algal cell consumption rate of 25100 and 22080 cells $Rotifer^{-1}$ day^{-1} respectively. In a similar way, Sarma and Nandini (2001) and Lucia-Pavon *et al.* (2001) reported highest production of *B. calyciflorus* (471±72 nos. mL^{-1}) when fed with *Chlorella* sp. at a dose of $4.5×10^6$ cells mL^{-1} compared to lower algal cell concentration applied at a dose of $0.5×10^6$ cells mL^{-1} (55.0±1.0 nos. mL^{-1}). Hotos (2002) reported that the higher growth of *Brachionus plicatilis* when fed with mixtures of *Chlorella* and *Asteromonas* at 93659 μm^3 $indi.^{-1}$ while Loka et al., (2016) reported high filtration rate ($12.2×10^{-5}$ cells mL^{-1} $indi.^{-1}$) and ingestion rate ($5.4x10^{-3}$ cells ml^{-1} ind^{-1} min^{-1}) in *B. plicatilis* when fed with *Nannochloropsis oculata* and *Isochrysis galbana*, respectively.

Abdel Rahman (2008) also reported highest count of *B. plicatilis* when fed with *N. oculata* at a density of 10402 cells $indi.^{-1}$ hrs^{-1} while Ashraf (2010) stated that the daily algal cells, *Chlorella vulgaris*, requirement of individual rotifer, *B. calyciflorus*, was approximately 33000-35000 cells day^{-1}. On the other hand, Bently et al., (2008) reported that much higher number of algal cells for individual rotifer not only affects environmental conditions but also reduces rotifer production. Nandini and Sarma (2000), Arimoro (2006) and Odo *et al.* (2015), reported that application of higher quantity of algal cell resulted in decreased growth rates of rotifers and lead to poor water quality caused by increased quantity of leftover food leading to over population in the system that can cause struggle for survivorship and environmental space (Sarma et al., 2001). These observations provided evidence that the higher production of rotifer, *B. calyciflorus*, can be achieved by applying *C. vulgaris* at optimum doses of 22000-25000 cells $rotifer^{-1}$ day^{-1} without affecting rotifer culture environment.

5.2.3.3 Effect of daily harvest rate on rotifer production

In the present study, daily harvesting at various rates (10, 30, 50 and 70%) had played significant effect on rotifers production ($p<0.05$). Daily harvesting at the rate of 30% of total culture volume maintained the rotifer culture at stationary phase (360.52±25.41 nos. mL^{-1}) without crash. Similarly, Sananurak et al., (2009) reported that the partial harvesting (20-30%) from a closed-recirculating, continuous culture system, for micro alga (*Tetraselmis suecica*) and rotifers (*Brachionus plicatilis*), maintained culture at stationary phase with $0.4×10^4$ cells mL^{-1} and 35 nos. mL^{-1}

production, respectively, for the culture period of 28 days. Suantika *et al.* (2000), Arimoro (2006) and Onal *et al.* (2015) also suggested that the partial harvest could (300-500%) help to maintain rotifer production and water quality in culture tanks. In contrast, Odo et al., (2015), suggested complete harvesting in batch and semi continuous culture systems on the 8th day instead of prolonging the culture which may lead to culture crash.

5.2.3.4 Effect of Recirculatory Aquaculture System (RAS) on rotifer production

Successful culture of fish and shrimp in various parts of the world is depending on live food sources (Yuan *et al.* 2018) and accordingly demand is also increasing progressively for rotifer, *B. calyciflorus,* in ornamental, food fish and shrimp hatcheries due to increasing awareness among fish farmers. Hence, there is a need to develop cost-effective, reliable, sustainable and simple system that can provide rotifers to fish producers throughout the season in required quantum with zero water exchange.

In commercial food fish aquaculture, rotifers are extensively cultured using baker's yeast or other yeast-based products in batch systems which are characterized by harvesting all rotifers once at a predetermined density level to minimize negative effects of poor water quality (Bentley et al., 2008 and Odo et al., 2015). In batch systems, rotifer densities rarely exceed 500-1000 nos. mL^{-1} (Suantika *et a.,* 2000) and higher densities are subject to sudden crashes which requires the use of backup systems thus increasing the production cost (Suantika *et a.,* 2003). Rotifer densities up to 2×10^4 and 1.6×10^5 nos. mL^{-1} have been reported from high-density and ultra-high density systems, respectively (Yoshimura et al., 2003). Still, unionized ammonia (NH_3-N), foam formation and proliferation of pathogenic bacteria are major problems in high-density cultures (Suantika et al., 2003) for which a composite and sustainable approach for their removal is required. For example, in a high density system, optimum water quality can be maintained by partial recirculation where culture water gets reconditioned through the series of filters.

In the present experiment, RAS based culture system had played significant role on rotifer production by maintaining optimum water quality throughout the culture with a daily production of around $5425\pm78.24\times10^5$ nos. $tank^{-1}$. The highest rotifer production was observed during summer (394.12 ± 15.41 nos. mL^{-1}) but was started declining gradually during monsoon (349.57 ± 25.44 nos. mL^{-1}) and winter (301.21 ± 14.33 nos. mL^{-1}) and again reached to its normal stage during summer (404.43 ± 24.33 nos. mL^{-1}). Similarly, Abu-Rezq *et al.* (1997) reported a chemostat culture system for rotifer production and showed daily output of 186×10^6 rotifers $1000L^{-1}$ tanks. A high-density (3000 nos. mL^{-1}) continuous recirculating system for rotifers (*B. rotundiformis*) fed on nonviable *Nannochloropsis oculata* and using sodium hydroxymethane sulfonate to neutralize ammonia was also developed by Bentley et al., (2008) with 125 L water holding capacity tanks for 30 days. Another high density rotifer production was reported in a closed recirculation system in 1 m^3 tank with a daily rotifer harvest of 2.1×10^9 rotifers $1000L^{-1}$ (Sauntika et al., 2003). Recently, Onal et al., (2015) reported small scale continuous culture system for rotifer, *B. plicatilis,* produced a mean of 12.3×10^5 rotifers day^{-1} with maximum rotifer density of 2188 nos. mL^{-1}.

RAS based culture system had various advantages over batch or semi-continuous culture practices which includes substantially shorter start-up time, less complicated, big tanks, more stable, simplified food source, more consistent harvest densities, simple and continuous harvest procedures, less contamination, control over the culture process, and much reduced need of water (Yoshumura et al., 1996; 1997; Suantika et al., 2000; Lim et al., 2003; Arimoro, 2006 and Onal et al., 2015).

The growth rate of rotifers has varied significantly according to seasonal variation, especially temperature. Similarly, Park (1998) also reported higher rotifer growth with temperature increase. Though dissolved oxygen (DO) was one of the limiting factors for rotifer growth (Park et al., 2001), no difference in rotifer quantity was observed as it was maintained always above 3.0 mg L^{-1}. In similar way, Park et al., (2001) reported higher rotifer growth when the DO level was maintained above 2.0 mg L^{-1}. Sananurak et al., (2009) also suggested for maintaining DO level always above 3.0 mg L^{-1} in semi-continuous, rotifer production. On the other hand, Rojo-Cebreros et al., (2017) has maintained DO level at 8.6-8.7 mg L^{-1} using respironics which supply pure oxygen for high density culture. Total hardness and alkalinity not showed any effect on rotifer concentration while pH was maintained between 7.4-8.8 as per reported by Suantika et al., (2000 and 2003) and Rojo-Cebreros et al., (2017) for getting higher production rate. Ammonia was a major threat in high density rotifer production, as it causes cessation in reproduction at 3-5 mg L^{-1} concentration. In the present study, ammonia concentration above 1.16 mg L^{-1} was not observed throughout the culture period. Similarly, in Suantika et al., (2000) report, daily harvest at 300-500% maintained ammonia level between 2.0-5.0 mg L^{-1}. Liang et al., (2018) also reported that more ammonia concentration had negative effect on reproductive performance, algal grazing rate, swimming speed and defensive phenotypes of *B. calyciflorus*. Liang et al., (2018) also suggested that, phenotypically, high concentrations of toxins lead to increase in length of posterolateral spine. The minimum level of ammonia throughout the culture may be accomplished due to incorporation of activated charcoal in sand filter. In similar way, Sichula et al., (2011) and Zeng et al., (2017) reported that ammonia can be removed from the aquaculture water using activated carbon.

5.3 Formulation of Probiotic incorporated encapsulated feed

5.3.1 Isolation, identification and characterization of *Lactobacillus sp.* from fish gut

In the present experiment, 11 colony numbers were observed from the gut of *D. dangila* while 09 colonies were observed in the gut of *P. sophore*. Total plate counts in *D. dangila* gut showed 1.7×10^7 cfu g^{-1} bacterial population while around 2.1×10^7 cfu g^{-1} bacterial population was observed in *P. sophore* gut. Similarly, Abraham and Banerjee (2007) and Uddin and Al-Harbi (2012) reported 1.84×10^6 cfu mL-1, 2.41×10^6 cfu mL^{-1}, 3.50×10^5 cfu mL^{-1} and 2.7×10^{10} cfu mL-1 bacterial population in *Labeo rohita, Cirrhinus mirigala, Cyprinus carpio* and *Clarias gariepinus* gut, respectively.

Ringo *et al.* (2006) reported that the population levels of adherent bacteria in foregut ($7 \times 10^{3-4}$ cfu g^{-1}), midgut (4×10^{3} cfu g^{-1}) and hindgut ($4.5 \times 10^{4-5}$ cfu g^{-1}) of Atlantic cod varied with different diet. In similar way, Hovda *et al.* (2007) determined the average bacterial counts in foregut (log 3.9 cfu g^{-1}), midgut (log 3.7 cfu g^{-1}) and hindgut (log 5.6 cfu g^{-1}) of Atlantic salmon (*Salmo salar*).

In case of *D. dangila* and *P. sophore*, 06 and 05, yellow colonies of lactic acid bacteria, respectively, were observed. According to the reports of Badis *et al.* (2004), Rengpipat *et al.* (2008) and Allameh *et al.* (2012), lactic acid bacteria (LAB) forms yellow colonies when the bromo-cresol purple was used as an indicator on MRS agar. The isolates were gram-positive, catalase and indole negative, mostly fermentative, non-motile and coccid to coccid rod shaped.

The antagonistic effect of isolates showed effective resistance against *Aeromonas hydrohila* by formation of either no zone or zone less than 3 mm. Similar results were reported by Ringo (1993) with 10% population of lactic acid bacteria in Atlantic salmon and Arctic charr (*Salvelinus alpinus*) gut. Similarly, Aly *et al.* (2008), Kaynar and Beyatli (2012) and Muthukumar and Kandipan (2015) reported bacterial resistance activity of lactic acid bacteria against *Aeromonas hydrohila,* with zone less than 5 mm. Kim and Austin (2008) also reported that the antibacterial ability of two probiotic strains from rainbow trout gut inhibited the growth of both *A. hydrophila and A. salmonicida* showing zone less than 5.00 mm.

The susceptibility and resistance pattern obtained with three isolates from each of *D. dangila* and *P. sophore* against different antibiotics showed less susceptibility to most of the antibiotics (no zone or zone less than 10 mm) (Bauer et al., 1966). In case of tetracycline, chloramphenicol, ofloxacin and ciprofloxacin all isolates from both the species showed zones less than 10 mm resulting lower resistance against them while in case of erythromycin, isolates from *D. dangila* showed less resistance (zones <10 mm) compared with *P. sophore* isolates. On the other hand, isolates from both the species showed more resistance against ampicilin, ceftazidime and penicillin. Resistance to specific antibiotics suggests that the probiotics can be used at the time when antibiotic treatment is required (Allameh et al., 2012; Das et al., 2014 and Muthukumar and Kandeepan, 2015). Similarly, Cebeci and Gurakan (2003) and Kim and Austin (2008) reported that the *Carnobacterium* strains showed resistance to antibiotics like ampicillin, gentamycin, kanamycin, streptomycin and penicillin but showed sensitivity against chloramphenicol, tetracycline and cotrimaxazole.

In the present investigation, pH (3-9) and bile salt (0.01-0.3%) concentration caused significant variation ($p<0.05$) in the growth of lactic acid bacteria. *Lactobacillu sp.* growth has been reduced considerably at low pH (pH - 3) for both the species, *D. dangila* (0.0042±0.0002) and *P. sophore* (0.0039±0.0004) at 600 nm optical density. On the other hand, highest growth was observed at neutral pH (pH - 7) for both the species, *D. dangila* (0.0160±0.0006) and *P. sophore* (0.0162±0.0007) at 600 nm optical density. Cebeci and Gurakan (2003) reported that *Lactobacillus plantarum* could survive at pH 4 but cannot multiply (Nguyen et al., 2007). Kim and Austin (2008) also reported that the highest growth of probiotic, *Carnobacterial* strains occurred at pH 5 to 10 isolated from rainbow trout gut. Allameh *et al.* (2012) also reported higher growth of *Lactobacillus mesenteroides* at neutral pH (pH - 7). Gatesoupe (1998), Rengpipat *et al.* (2008), Ringo and Balcazar *et al.* (2008) and Muthukumar and Kandeepan (2015), also reported that the lactic acid bacteria showed higher growth at neutral pH (pH - 7).

In the present study, bile salt concentration has showed significant difference in the growth of lactic acid bacteria ($p<0.05$). At 0.01% bile salt concentration, higher growth of *Lactobacillus sp.* was observed in both *D. dangila* (0.1250±0.0021 cfu mL^{-1}) and *P. sophore* (0.1290±0.0014 cfu mL^{-1}) at 600 nm optical density. Whereas, the growth was reduced considerably as bile salt concentration increased at 0.3% in both *D. dangila* (0.0183±0.0012 cfu mL^{-1}) and *P. sophore* (0.0163±0.0008 cfu mL^{-1}) at 600 nm optical density. Similar reports were given by Paludan *et al.* (1999), Salminen *et al.* (2004), Thapa *et al.* (2006) and Nguyen *et al.* (2007). According to Salminen *et al.* (2004), probiotic bacteria should tolerate bile salt concentration to grow and survive in fish intestine. Cebeci and Gurakan (2003) and Nguyen *et al.* (2007) reported that the *Lactobacillus plantarum* and *Lactobacillus fermentum* can survive in bile salt concentration ranging from 0 to 0.4% while Balcazar et al., 2008 reported that the *L. plantarum* isolated from rainbow trout gut can survive in 2.5% bile salt concentration for 1.5 hours without significant changes in viable counts.

5.3.2 Preparation and characterization of probiotic incorporated encapsulated feed

In the present experiment, diet of 50% crude protein incorporated with probiotic bacteria has showed no significant loss of lactic acid bacteria (LAB) when coated with sodium alginate (2.0%). When the feed was kept at a room temperature, around 15% loss was observed while at refrigerated condition (-20°C), less than 08% loss was recorded. Similarly, Goderska and Czarnecki (2008) reported that the highest numbers of live cells of *Bifidobacterium bifidum* (10^7 cfu g^{-1}) were observed in refrigerated condition after four months. Khalil and Mansour (1998), Kim *et al.* (2008) and Chavarri *et al.* (2012) also reported that the viability of free cells disappeared after 2 weeks, but encapsulated lactic acid bacteria like *Lactobacillus acidophilus, Bifidobacterium bifidum* and *Bifidobacterium infantis* survived well for 12 weeks. Adhikari *et al.* (2000) reported a significant difference between the viable counts of encapsulated *B. longum* compared to free cells at 4.4°C within 30 days. Abd-Talib et al., (2013) also reported loss of probiotic bacteria, *Lactobacillus sp.*, from 2.1×10^{10} cfu mL^{-1} to 7.82×10^{6} cfu mL^{-1} within two weeks of storage at 25°C.

The most commonly used bio gum for encapsulation of bacterial cells is sodium alginate due to non-toxicity, forms gentle matrices with calcium chloride to trap sensitive materials such as probiotic bacteria, the viability of bacteria is not affected during the encapsulated shelf life and the reversibility of immobilization as gels can be solubilized by sequestering calcium ions thus releasing the entrapped cells (Sheu and Marshall, 1993 and Shah and Ravula, 2000). Godward (2000) postulated that the encapsulation of *L. acidophilus* and *Bifidobacterium spp.* by sodium alginate enhanced the survival of probiotic bacteria. These studies clearly show that lactic acid probiotic bacteria are protected by sodium alginate encapsulation. According to Desmond *et al.* (2002), the level of moisture content should be at least 4% for prolonged storage of encapsulated lactic acid bacteria. In contrast, Godward (2000) observed that the free probiotic cells of *Lactobacillus acidophilus* and *Bifidobacterium infantis* survived better than encapsulated cells in Cheddar cheese and flourished that the cell death may be due to the exudation of organic acids within the calcium alginate.

5.3.3 Effect of probiotic incorporated encapsulated feed on larval growth of *D. dangila* and *P. sophore*

In the present experiment, probiotic incorporated encapsulated feed with different concentrations of *Lactobacillus sp.* (10^1 cfu g^{-1}, 10^3 cfu g^{-1}, 10^5 cfu g^{-1} and 10^7 cfu g^{-1}) has showed significant ($p< 0.05$) effect on larval growth, specific growth rate, average daily gain, feed conversion ratio and survival rate of both *D. dangila* and *P. sophore*. Incorporation of probiotic bacteria at a concentration of 10^5 cfu g^{-1} in feed showed highest final weight gain in both *D. dangila* (1.0233±0.033 g) and *P. sophore* (0.8233±0.003 g). On the other hand, the lowest weight gain for *D. dangila* (0.8733±0.048 g) and *P. sophore* (0.7467±0.023 g) was observed when the probiotic bacteria were incorporated at a rate of 10^7 cfu g^{-1} in feed. Similarly, Dharmaraj and Dhevendran (2010) and Arig *et al.* (2003) reported higher growth in *Xiphophorus helleri* and *Sparus aurata* when fed with lactic acid bacteria (LAB) incorporated feed. Similar reports were observed in *Penaeus indicus* (Ziaei-Nejad et al., 2006), *Cyprinus carpio* (Wang and Zirong 2006), *Litopenaeus vannamei* (Wang, 2007), *Paralichthys olivaceus* (Ye et al., 2011), *Labeo rohita* (Mohapatra et al., 2012) and *Epinephelus coioides* (Sun et al., 2013). Indeed, Merrifield *et al.* (2010) reported that high intestinal levels of *Bacillus subtilis* and *Bacillus licheniformis* (>80 %) caused deformed alevins with reduced growth performance and feed utilization in rainbow trout.

The specific growth rate and average daily gain also showed significant variation in respond to the concentration of probiotic bacteria incorporated. The highest specific growth rate and average daily gain observed for *D. dangila* were 0.1324±0.0023% and 0.0334±0.0001 g day^{-1}, respectively, whereas, in *P. sophore* it was 0.1384±0.0015% and 0.0270±0.0001 g day^{-1}, respectively, at 10^5 cfu g^{-1} lactic acid bacteria incorporation. Similarly, Ziaei-Nead *et al.* (2006) reported effects of *Bacillus spp.* on different developmental stages of *F. indicus* showing significant difference in the growth rate compared to control groups. Dohail *et al.* (2009) evaluated the effects of *Lactobacillus acidophilus* on the growth performance in African catfish, *Clarias gariepinus* fingerling showing significant elevation in the growth performance parameters like specific growth rate, relative growth rate, protein efficiency ratio, feed conversion ratio and survival rates in comparison with the control.

Consequently, the highest survival rate for *D. dangila* (78.41±0.84%) and *P. sophore* (90.08±0.93%) with efficient feed conversion ratio (1.51±0.19; 1.54±0.02, respectively) was observed at 10^5 cfu g^{-1} lactic acid bacteria incorporation. Similarly, Ghosh *et al.* (2008) reported increase in growth and survival of *Xiphophorus heleri* and *Poecilia reticulate* when intestinal isolate of *Bacillus subtilis* from *Cirrhinus mrigala* were incorporated into their diet for 50 and 90 days. Ghosh *et al.* (2007) reported that the strain of *Lactobacillus spp.* isolated from intestine of *Cirrhinus mrigala*, incorporated in diet at a concentration of 10^6 cfu g^{-1} showed increase in gonadosomatic index, fecundity, viability and fry production capacity of the females of *Poecilia reticulata, P. sphenops, Xiphophorus helleri,* and *X. maculates,* within one year. Similar findings are well noted by Avella *et al.* (2010) who fed the mixture of four strains of *Lactobacillus sp.* to sea bream, *Sparus aurata* larvae and juvenile through feed showed significant increase in the growth parameters, feeding efficiency and metabolism.

Histological analyses showed significant difference ($p<0.05$) in number of villi and villi length formed across anterior and posterior margin of intestine and muscular thickness in both *D. dangila* and *P. sophore*. Villi are specialized tissues for the absorption of nutrients in the intestine (Ferraris et al., 1990 and Oxley et al., 2007). Therefore, the maximum numbers of villi with more surface area are required for the maximum absorption of nutrients from food particles efficiently (Bakke et al., 2011).The average number of villi and villi length were increased in both *D. dangila* (34.11±4.21; 214.41±14.25 μm, respectively) and *P. sophore* (36.52±21.12; 198.52±15.56 μm, respectively) significantly than control. The muscular thickness was reduced significantly in *D. dangila* intestine (32.41±2.71 μm) when compared with control (34.11±2.14 μm) but significant difference in muscular thickness was not observed in *P. sophore* in treatment (29.20±6.14 μm) and control group (30.41±1.34 μm). Further the histological sectioning confirmed that the intestinal villi get damaged showing abnormal structures with superfluous mucus secretion at higher dosage of *Lactobacillus sp.* in *D. dangila,* in contrast to the normal structures were observed in *P. sophore* intestinal sections. Similar reports were postulated by Morrison (1987), Santamaría, *et al.* (2004) and Falk-Petersen, (2005) in larva of Salmonids and barbs. In similar way, several reports revealed that the extreme dose of bacterial mixture to larvae resulted in reduction of intestinal epithelial cells height and the number of villi (Kostomarova, 1962; Ehrlich et al., 1976; Theilacker and Watanabe, 1989 and Gisbert et al., 2004). Firdaus-Nawi et al., (2013) reported that the thickness of lamina propria was inversely proportional with the number of villi in *Epinephelus fuscoguttatus*.

5.4 Larval rearing of *Danio dangila* and *Puntius sophore*

5.4.1 Effect of rearing system and stocking density on *Danio dangila*

The fish reared in nursery raceway system significantly ($p<0.05$) exhibited higher survival rate (62.86%) compared to FRP tanks based rearing system. In the case of stocking density, significantly higher survival rate was observed in 30 nos. L^{-1} stocking density (81.00%) which was almost equal for stocking density at the rate of 10 nos. L^{-1} (80.55%) with significant interaction effect ($p = 0.041$). The final weight was significantly ($p<0.05$) higher in fishes reared in nursery raceways (1.229 g) than FRP tanks (1.078 g) ($p = 0.043$) while the higher mean weight was observed in fish larva stocked at 30 nos. L^{-1} (1.323 g) ($p = 0.038$). In case of stocking density, almost equal average daily gain was observed at 10 nos. L^{-1} (0.040 g) and 30 nos. L^{-1} (0.042 g) with no significant difference. In case of specific growth rate, fishes reared in nursery raceways (0.132%) showed significantly higher specific growth rate ($p<0.05$) while in comparison with stocking densities, significantly higher specific growth rate was observed at 10 nos. L^{-1} and 30 nos. L^{-1} stocking densities (0.134%).

Similarly, studies conducted on *Xiphophorus heleri* and guppies, Poeciliidae showed similar results with respect to stocking densities and water circulation (Vine and Kaiser 1996; Oliver and Kaiser 1997 and Ako et al., 2005). In outdoor ponds or traditional open culture systems, fishes are grown at a very low stocking ratio (0.2–0.3 fish L^{-1}) (Fernando and Phang 1985; Degani 1993; Jha and Barat 2005; Jha 2010 and Jahedi et al., 2012). The optimal stocking density (3 fish L^{-1}) for the guppy in RAS was first demonstrated by Oliver and Kaiser (1997).

5.4.1 Effect of rearing system and stocking density on *Puntius sophore*

The fish reared in nursery raceway system significantly ($p<0.05$) exhibited higher survival rate (76.30%) compared to FRP tanks (68.06%). In the case of stocking density, significantly higher and almost equal survival rate was observed in 30 nos. L^{-1} (91.79%) and 10 nos. L^{-1} (91.72%) stocking densities at significant interaction effect (p = 0.00). The final weight was significantly ($p<0.05$) higher in fishes reared in nursery raceways (1.062 g) than FRP tanks (0.920 g). By showing significant interaction effect of rearing system and stocking density (p = 0.00), the higher mean weight was observed in fishes stocked at 10 nos. L^{-1} (1.097g) followed by 30 nos. L^{-1} (1.022 g) with no significant difference between treatments. Higher average daily gain (0.034 g day^{-1}) was observed in nursery raceway based system followed by FRP tanks (0.029 g day^{-1}) significantly ($p<0.05$). In case of stocking density, highest average daily gain was observed at 10 nos. L^{-1} (0.0349 g day^{-1}) followed by 30 nos. L^{-1} (0.0325 g day^{-1}). In case of specific growth rate, fishes reared in nursery raceways (0.138%) showed significantly higher specific growth rate ($p<0.05$) followed by rearing in FRP tanks (0.133%). In comparison with stocking density, higher specific growth rate was observed at 10 nos. L^{-1} stocking density (0.141%) and 30 nos. L^{-1} (0.140%) compared to higher stocking densities.

Similarly, Irwin *et al.* (1999) studied the effect of stocking density (0.7, 1.1, 1.5 and 1.8 kg m^{-2}) on the growth of turbot, for 45 days, and found that stocking density inversely affects growth rate and mean weight. In many cultured fish species, growth is inversely related to stocking density and this is mainly attributed to social interactions (Holm et al., 1990; Haylor, 1991; Bjornsson, 1994; Huang and Chiu, 1997; Irwin et al., 1999 and Ma et al., 2006). Iglesias et al., (1987) and Imsland et al., (2003) reported that the turbot can be reared at very high stocking densities in commercial operations systems like raceways with stocking densities generally range from 25 to 30 kg m^{-3} with a maximum reported sustainable density of 75 kg m^{-3}.

In the present experiment, incidence of mortality at higher stocking densities recorded which could be due to behavioral changes like rapid spread of virulent pathogens, stress, as a result of competition for food and space, poor water quality, poor handling and deterioration of health due to positive interaction of stressful factors (Barton and Iwama, 1991; Cruz and Ridha, 1991; Vera Cruz and Mair, 1994; Akinwole et al., 2014 and Ronald et al., 2014).

The mean survival rates of around 89.4%-98.5% were reported by Aksungur et al., (2007) in juvenile turbot when stocked at 60 fish m^{-2} which was supported by similar studies conducted by Heap and Thorpe (1987), Person-Le Ruyet *et al.*

(1991) and Sunde *et al.* (1998). Similarly, the survival rate declined significantly in the highest stocking density (90 fish m^{-2}) probably due to high stocking rate and low space. In similar way, Tibile *et al.* (2016) reported that the lower densities of 1 juvenile 10 L^{-1} and 1 juvenile 7.5 L^{-1} showed significantly higher growth and survival in Discus fish, *Symphysodon aequifasciatus*.

CHAPTER – 6

Summary and Conclusion

The present study deals with 1) Induced breeding of indigenous ornamental fishes, *Danio dangila* and *Puntius sophore*, from North-East India, 2) Mass production of freshwater live food organism, Rotifer, *Brachionus calyciflorus*, and 3) Formulation of probiotic incorporated encapsulated feed 4) Larval rearing in advanced aquaculture systems like raceways with a view to develop standard breeding and larval rearing techniques in nursery raceways with highest possible stocking densities. The important findings are summarized as follows:

1. In the present investigation, *Danio dangila* and *Puntius sophore* were bred successfully under captive condition with the administration of molecular analogue based synthetic hormone, WOVA-FH. *Danio dangila* administered with a dose of 0.5 mL kg^{-1} for female and 0.3 mL kg^{-1} for male and *Puntius sophore* administered with a dose of 0.4 mL kg^{-1} for female and 0.2 mL kg^{-1} for male yielded highest egg output (1002.50±52.6; 881.50±85.8), fertilization rate (84.0±2.72%, 78.50±2.72%) and hatching rate (86.5±2.9%, 83.75±2.39%), respectively, within optimum latency period (6.00 – 8.00 hours) for both the species.
2. Mass production of freshwater rotifer, *Brachionus calyciflorus*, in outdoor culture system was accomplished by using *Chlorella vulgaris* as a preferred food type, with highest rotifer production (180.33±2.18 nos. mL^{-1}), than other food types. For increasing the Rotifers (*B. calyciflorus*) production, quality and quantity of food type also played significant impact. Nutrient rich *C. vulgaris* cells fed at the rate of 3 times day^{-1} to the rotifers, resulted highest rotifer production (256.33±3.8 nos. mL^{-1}) than the tanks fed at 2 times per day (186.67±1.6 nos. mL^{-1}) on 8th day onwards and could be maintained for further days by partial harvesting of the rotifers.
3. The present investigation propose that the algal growth, protein, chlorophyll a, chlorophyll b, astaxanthin amounts and cell size of *C. vulgaris* could be regulated by different light wavelengths. The cell multiplication ($65.220 \pm 0.17 \times 10^6$ cells mL^{-1}), chlorophyll a (0.177±0.17 mg mL^{-1}) and chlorophyll b (0.141±0.13mg mL^{-1}) contents were significantly increased in green light illumination while red light illumination played important role in increase of protein (0.132±0.14 mg mL^{-1}) and astaxanthine production (0.0126±0.08 mg mL^{-1}). For enhancement of lipid content (0.252±0.24

mg mL^{-1}), white illumination was found to be effective. In addition, the present study devised easy and safe method for reducing frequent crashes in algal culture tanks due to ciliate contamination with an application of common purified salt at the rate of 60 g L^{-1}, which effectively controlled ciliate growth within 24 hours from algal culture tanks.

4. 4. For further up-scaling, mass production of freshwater rotifer was demonstrated in the RAS based culture system for continuous production of rotifers without crash and contamination by identifying ideal food type (*C. vulgaris*), minimum algal cell requirement for a rotifer in a day (22000-25000 cells $rotifer^{-1}$ day^{-1}) and standard harvesting rate (30% day^{-1}) for maintaining the culture at its stationary phase with uniform rotifer production (360.52±150.41 nos. mL^{-1}) throughout the season with a daily production of around $5425\pm78.24\times10^5$ nos. $tank^{-1}$, significantly.
5. 5. Probiotic bacteria (*Lactobacillus sp.*) isolated from fish intestine was produced in mass and encapsulated at different concentrations (10^1 cfu g^{-1}, 10^3 cfu g^{-1}, 10^5 cfu g^{-1} and 10^7 cfu g^{-1}) in fish feed. For verification of viability of lactic acid bacteria, diet of 50% crude protein incorporated with probiotic bacteria was analyzed which showed no significant loss of lactic acid bacteria (LAB) when encapsulated with sodium alginate (2.0%). When the feed was kept at room temperature, around 15% loss was observed while kept at refrigerated condition (-20°C), less than 08% loss was recorded. In the larval rearing experiment, the highest final weight gain was observed in both the species, *D. dangila* (1.0233±0.033 g) and *P. sophore* (0.8233±0.003 g), when the probiotic bacteria incorporated encapsulated feed were fed at a concentration of 10^5 cfu g^{-1} with maximum average daily gain (0.0334±0.0001 g day^{-1} and 0.0270±0.0001 g day^{-1}), respectively. In the histological viewpoint, the fish gut health was analyzed and showed that the average number of villi and villi length was increased in both *D. dangila* (34.11±4.21; 214.41±14.25 μm) and *P. sophore* (36.52±21.12; 198.52±15.56 μm) significantly.
6. 6. In aquaculture, the term raceway is a highly generic and implies little more than a water impoundment with water flow-through and Recirculatory Aquaculture system (RAS). Raceway farming would enable the aquaculture sector to grow at a different level in our country. In the present investigation, larval rearing was done in two different rearing systems *viz.* FRP tanks and RAS based nursery raceways. Fish larvae of *D. dangila* and *P. sophore*, stocked at the rate of 30 nos. L^{-1} in recirculatory aquaculture system (RAS) based nursery raceways, showed highest survival (81.00% and 91.79±1.67%), final weight gain (1.323 g and 1.022±0.029 g) and highest specific growth rate (0.134±0.001% day^{-1} and 0.140±0.001% day^{-1}), respectively, compared to FRP tanks based rearing system.

To conclude, the present study demonstrated the breeding technique of two North-East region based ornamental fishes namely, *Danio dangila* and *Puntius sophore*, and provided insight into embryonic development stages. For initial feeding of fish larvae, mass production techniques of freshwater live food organism, Rotifer (*Brachionus calyciflorus*), was standardized in both outdoor and RAS based indoor culture systems in a continuous culture mode. Further, probiotic incorporated encapsulated feed was produced for enhancing larval health. For larval rearing, the study demonstrated maximum production in recirculatory aquaculture system (RAS) based nursery raceways with highest possible stocking density and with minimum water usage. The results of this investigation shall be useful for fish breeders and ornamental fish farmers for expanding aquariculture, species restoration and conservation.

CHAPTER – 7

References

Abareethan, M. and Amsath, A., 2015. Characterization and evaluation of probiotic fish feed. *International Journal of Pure and Applied Zoology*. 3 (2), 148-153.

Abdel Rahman, S.H, Abdel Razek, F. A., Abou Zeid, A. E., Mohamed, A.F., 2008. Population growth rate, fecundity, filtration and ingestion rate of marine rotifer *Brachionus plicatilis* fed with motile and immotile microalgae. *Egyptian Journal of Aquatic Research*. 34, 426-439.

Abidi, S.A.H., Thakur, N.K., 1997. Advances in keeping and breeding ornamental fishes. *Training manual*. Central Institute of Fisheries Education, Versova, Mumbai.

Abou-Shanab, R.A.I., Manjinder Singh, Anangelica Rivera-Cruz, Grace Power, Thomas Bagby-Moon, Keshav Das, 2016. Effect of *Brachionus rubens* on the growth characteristics of various species of microalgae. *Electronic Journal of Biotechnology*. 22, 68-74.

Abu-Rezq, T.S., Al-Shimmari, J., Dias, P., 1997. Live food production using batch culture and chemostat systems in Kuwait. *Hydrobiologia*. 358, 173-178.

Ajah, P.O., 2008. Growth characteristics of the monogonont rotifer, *Asplanchna priodonta* Gosse 1850 on three algae species. *Turkish Journal of Fisheries and Aquatic Sciences*. 8, 275-282.

Ajah, P.O., 2010. Mass culture of rotifer (*Branchionus quadridentatus* (Hermainn, 1783) using three different algal species. *African Journal of Food Science*. 4, 8-8.

Aksungur, N., Aksungur, M., Akbulut, B., Kutlu, I., 2007. Effects of stocking density on growth performance, survival and food conversion ratio of turbot (*Psetta maxima*) in the net cages on the Southeastern coast of the Black Sea. *Turkish Journal of Fisheries and Aquatic Sciences*, 7, 147-152.

Akteruzzaman, M., Kohinoor, A.M., Hussain, M.G, Shah, M.S., 1991. Observation on the induced breeding of *Mystus cavasius* (Ham) in Bangladesh. *Bangladesh Journal of Fisheries*. 14, 101-105.

Alam, M.J., 2006. BFRI Scientists succeeded to breed a brackishwater catfish, *Mystus gulio* in Captivity. *In:* Fisheries Newsletter, edited by Mazid, M. A. Published by Bangladesh Fisheries Research Institute. 14 (1), 1-18.

Alikuahi, K.H., Vijaya, Lakshmanan, M.A., Ibrabim, K.H, 1960. Preliminary observations on the spawning of Indian carps, induced by injection of pituitary hormones, *Indian Journal of Fisheries*. 7 (1), 1-19.

Al-Qasmi, M., Raut, N., Talebi, S., Al-Rajhi, S., Al-Barwani, T., 2012. A review of effect of light on microalgae growth. *Proceedings of the World Congress on Engineering.* 4-6.

Aly, A.H., Edrada-Ebel, R.A., Wray, V., Müller, W.E.G., Kozitskaya, S., Hentschel, U., Proksch, P. and Ebel R., 2008. Bioactive metabolites from the endophytic fungus *Ampelomyces* sp. isolated from the medicinal plant *Urospermum picroides*. *Phytochemistry.* 69, 1716-1725.

Anderson, J.W., 1962. More about Fancy Goldfish. *The Aquarium.* 31 (6), 310-315.

Anderson, J.W., 1963. Gold Fish Forming – a science, an art and an industry, *Ibid.* 32 (8), 7-12.

Anderson, J.W., 1965. Gold Fish can be bred indoors II, *Ibid.* 32 (5), 15-17.

Andrew, C., 1986. The ornamental fish trade and fish conservation. *Journal of Fish Biology.* 37 (A), 53-59.

Angami, V.K., 2012. Studies on the bionomics and breeding biology of *Danio dangila* (Hamilton, 1822) and *Puntius chola* (Hamilton, 1822), two Rheophilic ornamental fishes of Nagaland. PhD. Thesis. Retrieved from http://localhost: 6060/ xmlui/ handle/1/116.

Angami, V.K., Ahmed, S.U., 2010. Length-weight relationship of *Danio dangila* (Ham.) from the drainages of Nagaland. *Life Science Bulletin.* 7 (1), 59-61.

Arimoro, F.O., 2006. Culture of the freshwater rotifer, *Brachionus calyciflorus*, and its application in fish larviculture technology. *African Journal of Biotechnology.* 5 (7), 536-541.

Ashraf, M., Ullah, S., Rashid, T., Ayub, M., Bhatti, E.M., Naqvi, S.A., Javaid, M., 2010. Optimization of indoor production of fresh water rotifer, *Brachionus calyciflorus,* a preliminary study. *International Journal of Agricultural Biology.* 12, 719-723.

Azad, I.S., Shimray, D.K., 1991. First success in induced breeding of Indian and exotic carps in Manipur using ovaprim-C. *Fishing Chimes.* 10, 28-29.

Barua, G., Mollah, M.F.A., 1987. Observation on the spawning behaviour of *Mustus tengara* (Ham.) *Bangladesh Journal of Fisheries.* 10, 43-46.

Battaglene, S.C., Mcbride, S., Talbot, R.C., 1994. Swim bladder inflation in larvae of cultured sand whiting, *Sillago ciliata* Cuvier (Sillaginidae). *Aquaculture.* 128, 177-192.

Bauer, J.D., 1990. *Clinical Laboratory Methods,* 9th edition, pp. 182-183.

Begum, M., Pal, H.K., Islam, M.A., Alam, M.J., 2009. Embryonic and larval development of *Mystus gulio* (Ham.). *Bangladesh Journal of Fisheries Research.* 13 (2), 169-177.

Behera, B.K., Das, P., Singh, N.S., Sahu, A.K. 2007. Observation on the induced breeding of *Labeo bata* (Hamilton) with Ovaprim and Ovatide as inducing agents with a note to its development. *Journal of Aquaculture.* 15, 11-17.

Behera, B.K., Das, P., Singh, N.S., Sahu, A.K., 2009. Breeding and seed production of *Labeo gonius* (Hamilton) under captive condition in Manipur. *Journal of Aquaculture.* 17, 1-6.

Behera, B.K., Das, P., Singh, N.S., Sahu, A.K., 2010. Captive breeding of an endemic medium carp Pengba, *Osteobrama belangeri* (Val.) with Wova-FH in Manipur. *Journal of Aquaculture.* 18, 23-29.

Bentley, C.D., Carrol, P.M., Watanabe, W.O., Riedel, A.M., 2008. Intensive rotifer production in a pilot-scale continuous re-circulating system using non-viable microalgae and an ammonia neutralizer. *Journal of World Aquaculture Society.* 39, 625-635.

Bhattacharjya, B.K, Choudhury, M., 2004. Ornamental fish culture and trade in North Eastern India. *Workshop Proceeding,* CIFRI, Kolkata, pp. 1-14

Bhattacharjya, B.K, Sarma, S., Sarma, P., 2004. Collection, conditioning and transportation of indigenous ornamental fishes with special reference to those of the Northeast. *In*: Ornamental fish culture and trade in Northeastern India (Eds. B.K. Bhattacharjya and M. Choudhury), 2004. *Workshop Proc., CIFRI, Barrackpore, Kolkata.* 48-58.

Bhattacharya, S., Dasgupta, S., Datta, M., Basu, D., 2002. Biotechnology Input in Fish Breeding. *Indian Journal of Biotechnology.* 1, 29-38.

Bhuiyan, A.S., Islam, K., Zaman, T., 2006. Fecundity and ovarian characteristics of *Puntius gonionotus* (bloch/bleeker) (cyprinidae: cypriniformes). *Journal of Biological Science.* 14, 99-102.

Bijoya K., Bishal D., Sankar Kumar G., 2014. Jiribam, the Ornamental Fishes' Hot Spot Zone of Manipur, India. *IOSR Journal of Agriculture and Veterinary Science.* 7 (1), 85-91.

Billard, R., Alagarswami, K., Peter, R.E., Breton, B., 1983. Potentialisation par le pimozide des effets du LHRH-A sur la secretion gonadotrope hypophysaire, ovulation et la spermiation chez la carpe commune, *Cyprinus carpio. Comptes Rendus Academie des Sciences Paris.* 296, 181-184.

Billiard, R., Bieniarz, K., Peter, R.E., Sokolowska, M., Weil, C., 1984. Effect of LHRH and LHRH-a on plasma GtH levels and maturation/ovulation in common carp, *Cyprinus carpio* kept under various environmental conditions. *Aquaculture.* 41, 245-254.

Bindu, L., Aminal, S., Linsha, R., Sreelekshmi, G., Archna, P.L., 2014. Breeding and early development of the nest building gourami *Trichogaster trichopterus. International Journal of Research in Fisheries and Aquaculture.* 4 (4), 175-180.

Biswas, S.P., Santosh, K.S.A., Das J.N., 2015. Conservation and Management of Ornamental Fish Resources of North East India. *Journal of Aquaculture Research and Development.* 6 (3) 1-3.

Blair, M.F., Kokabian, B., Gude, V.G., 2014. Light and growth medium effect on *Chlorella vulgaris* biomass production. *Journal of Environmental Chemical Engineering.* 2, 665-674.

Bligh, E.G., Dyer, W.J., 1959. A rapid method of total lipid extraction and purification. *Canadian Journal of Biochemistry and Physiology*. 37 (8), 911-917.

Borch K., Ine E.P., Raimo O. H., 2015. The use of probiotics in fish feed for intensive aquaculture to promote healthy guts. *Advances in Aquaculture and Fisheries Management*. 3 (7), 264-273.

Borch, K., Pederson, I.E. and Hogmo, R.O., 2015. The use of probiotics in fish feed for intensive aquaculture to promote healthy guts. *Advances in Aquaculture and Fisheries Management*. 3 (7), 264-273.

Boussiba, S., Vonshak, A., 1991. Astaxanthin accumulation in the green alga Haematococcus pluvialis. *Plant Cell Physiology*. 32, 1077-1082.

Bryant, P.L., Matty, A.J., 1980. Optimization of Artemia feeding rate for carp (*Cyprinus carpio*) larvae. *Aquaculture*. 21, 203-212.

Cacot, P., 1999. Etude du cycle sexuel et maitrise de la reproductionde Pangasius bocourti. (Sauvage, 1880) et *Pangasius hypophthalmus* (Sauvage, 1878) dans le delta due Mekong auViet-Nam. PhD Thesis, CanTho University, Vietnam.

Carvalho, A.P., Silva, S.O., Baptista, J.M., *et al.*, 2011. Light requirements in microalgal photobioreactors, an overview of biophotonic aspects. *Applied Microbiology and Biotechnology*. 89 (5), 1275-1288.

Castell, J., Blair, T., Neil, S., Howes, K., Mercer, S., Reid, J., Gullison, B., Dhert, P., Sorgeloos, P., 2003. The effect of different HUFA enrichment emulsions on the nutritional value of rotifers (*B. plicatilis*) fed to larvae haddock (*Melanogrammus aeglefinus*). *Aquaculture International*. 11 (1-2), 109-117.

Chaco, P.I., Kuriyan, G.K., 1984. On the food of nine species of Puntius of Madras. *Scientific Culture*. 13 (8), 387-391.

Chatterjee, N.R., Samiran P., Talwar, N.A., 2013. Induced breeding of fresh water Angelfish (*Pterophyllum Scalare*) using ova prim. *Journal of Agriculture and Veterinary Science*. 3 (3), 24-28.

Chaudhuri, H., 1962. Breeding of *Puntius sarana* (Hamilton) and observations on its life history and bionomics. *Proceedings on Indian Science Congress*. 49 (3), 390-395.

Chaudhuri, H., Alikunhi, K.H., 1957. Observations on the spawning in Indian major carps by hormone injection. *Current Science*. 26 (12), 381-382.

Chaudhury, M., Biswas, S.P., 2003. Prospects of commercial rearing of a Wetland Ornamental Fish *Chana Barca* in Assam *In:* Participatory Approach for Fish Biodiversity Conservation in North East India (eds) Mahanta, P.C. and Tyagi, K.K. NBFGR, Lucknow: 231–232.

Chavarri M., Maranon I., Ares R., Ibanez F.C., Marzo F., Villaran M.C., 2010. Microencapsulation of a probiotic and prebiotic in alginate-chitosan capsules improves survival in simulated gastro-intestinal conditions. *International Journal of Food Microbiology*, 142, 185-199.

Chew, W.Y.S., Lim, H.S., 2006. Some improvement to rotifers (*Branchionus rotiundiformis*) mass culture method. *Singapore Journal of Primary Industry*. 32, 52-58.

Cho, S.H., Hur, S.B., Jo, J.Y., 2001. Effect of enriched live feeds on the survival and growth rates in the larval Korean rockfish, *Sebates schlegeli* Hilgendorf. *Aquaculture Research.* 32 (3), 199-200.

Choi, K., Jeon, S. Kim, I., Son, Y., 1990. *Silurus asotus. In:* K. Choi, S. Jeon, I. Kim, and Y. Son, editors. Coloured illustrations of the freshwater fishes of Korea. Hyang-munsa, Seoul, Korea. Pp. 152–153.

Choudhuri, H., Alikunhi, K.H., 1957. Observations on the spawning in Indian major carps by hormone injection. *Current Science.* 26 (12), 381-382.

Clyde, S., Tamaru, C., Richard, B., Brown, C., 1997. Manual for commercial production of Tiger Barb, *Copoetra tetrazona* temporary pair tank spawner. *Center for Tropical Subtropical Aquaculture.* 129, 1-48.

Convention on Biological Diversity, 2006. Global Biodiversity Outlook 2, *Secretariat of the Convention on Biological Diversity,* Montreal, 7: 81 pp.

Craig, S.R., Connie, R., Holt, J.G., 1994. The effects of enriching live foods with unsaturated fatty acids on the growth and fatty acids on the growth and fatty acid composition of larval red drum *Sciannops ocellatus. Journal of World Aquaculture Society.* 25 (3), 424-434.

Crim, L.W., Peter, R.E., Van Der Kraak, G.V., 1987. The use of LHRH analogues in aquaculture. *In:* B.H. Vickery and JJ. Nestor Jr, editors. *LHRH and its analog: Contraceptive and therapeutic applications,* part 2, MTP Press, Boston, MA. Pp. 489-498.

Dahanukar, N., 2010. *Puntius sophore.* The IUCN Red List of Threatened Species 2010, e.T166623A6249514. http,//dx.doi.org/10.2305/IUCN.UK. 2010-14. RLTS.T166623A6249514.en.

Daliry, S., Hallajisani, A., Mohammadi, R,J., Nouri, H., Golzary, A., 2017. Investigation of optimal condition for *Chlorella vulgaris* microalgae growth. *Global Journal of Environmental Science and Management.* 3 (2), 217-230.

Das, M., 1989. Studies on the systematics and Bioecology of the Ichthyospecies belonging to the Genus Tor from the Jia Bharali drainage of Assam (India), *Ph. D. Thesis,* Gauhati University, 15, pp. 269.

Das, P.B.K., Behera, D.K., Meena, S.K., Singh, S.C., Mandal, S., Sahoo, Das, A.K., 2016. Comparative efficacy of different inducing agents on breeding performance of a near threatened cyprinid *Osteobrama belangeri* in captivity. *Aquaculture Reports.* 4, 178-182.

Das, S.K., Bhattacharya, B.K., Sarma, K., 1994. Induced spawning and hatching of Tawes, *Puntius javanicus* (Bleeker). *Asian Fisheries Science.* 7, 191-194.

David, A., Rahman, M.F., 1975. Studies on some aspects of feeding and breeding of *Puntius pulchallus* (Day) and its utility in cultural waters. *Journal of Inland Fisheries society of India.* 7, 225-238.

Davies, B.H., 1976. Analysis of carotenoid pigments. *In,* Chemistry and Biochemistry of Plant pigments. Godwin, T.W. (ed.) Academic Pres, New York. 2, 38-165.

De Pauw, N., Van Vaerenbergh, E., 1983. Microalgal wastewater treatment-systems: potentials and limits, *In:* Convegno Internazionale su "Fitodepurazione e impiego delle biomasse prodotte". pp. 211-28.

de-Bashan, L.E., Antoun, H. and Bashan, Y., 2005. Cultivation factors and population size control uptake of nitrogen by the microalgae *Chlorella vulgaris* when interacting with the microalgae growth- promoting bacterium *A. brasilense. FEMS Microbiology and Ecology,* 54, 197-203.

Degani, G., 1993. Growth and body composition of juvenile *Pterophyllum scalare* (Lichtonstein) (Pisces:Cichlidae) at different densities and diets. *Aquaculture and Fishery Management.* 24, 725-730.

Deng, W., Dong, X.F., Tong, J.M., Zhang, Q., 2012. The probiotic *Bacillus licheniformis* ameliorates heat stress-induced impairment of egg production, gut morphology, and intestinal mucosal immunity in laying hens. *Poultry Science.* 91, 575-582.

Devaraja, S., Bharath, M., Deepak, K., Suganya, B., Vishal, B.S., Swaminathan, D., Meyyappan ,N., 2017. Studies on the Effect of Red, Blue and White LED Lights on the Productivity of *Chlorella Vulgaris* to Treat Dye Industry Effluent. *Advances in Biotechnology and Microbiology.* 6 (2), 1-6.

Dey, S.C, Kakati, M., Sarmah, S.K., 2002. Pre-investment feasibility study on Ornamental fish in the North Eastern Region, *Final Report* NEDFi, Guwahati – 781 005.

Dey, S.C., 1973. Studies on the distribution and taxonomy of the ichthyofauna of the hill streams of Kamrup-Khasi-Garo regions of Assam with special reference to the functional morphology of some rheophilic fishes, D. Sc. Thesis, Calcutta University, xi + 299 p.

Dey, S.C., 1975. Ecomorphology of the ichthyofauna of river Pagladia within Kamrup district. *Journal of Assam Science and Society.* 18 (1), 1-5.

Dey, S.C., 1982. A critical analysis on the fish and fisheries of Assam, *Proceedings of All India Seminar on Ichthyology.* 3, 16.

Dey, S.C., 1984. Report of ornamental fish trade in North East India, keynote address, National workshop on ornamental fish, Guwahati.

Dey, S.C., 1995. Ornamental and endangered fishes of North East Region, *Assam Science Society*, Gauhati University. 3, 16.

Dey, S.C., Sarmah, S.K., 2000. Breeding lineament of *Glossogobius giuris* (Hamilton, 1822), *Regl. Sem. On recent trends in Zoology,* NEHU. 1-12.

Dey, S.C., Sarmah, S.K., 2003. Ornamental Fish paragon of North Eastern India: *In*: *Zoonica,* A collection of Lectures delivered in the Department of Zoology, Pragjyotish College, Guwahati, p 9–12.

Dey, V.K., 2016. The global trade in ornamental fish. *INFOFISH International.* 4, 52-55.

Dhar, B., Ghosh, S.K. 2015. Identifying Ornamental Fishes of North east India Through DNA. *Barcoding.* www.researchgate.net/publication/277078233

Dhert, P., 1998. Rotifer, *In*. P. Lavens and P. Sorgeloos. *Manual on the production and use of live food for aquaculture*. Laboratory of Aquaculture and *Artemia* Reference. pp, 49-76 . Center, University of Ghent Belgium. F.A.O. Fisheries Technical paper No. 361, 305.

Dhert, P., Rombaut, G., Suantika, G., Sorgeloos, P., 2001. Advancement of rotifer culture and manipulation techniques in Europe. *Aquaculture*. 201, 129-146.

Dixit, R.K., Agarwal, N., 1974. Studies on the developmental rhythm in the oocyte of *Puntius cophose* (Ham.) *Acta. Anat*, 90, 133–144.

Drori, S., Ofir, M., Levavi-Sivan, B., Yaron, Z., 1994. Spawning induction in common carp (*Cyprinus carpio*) using pituitary extract or GnRH superactive analogue combined with metoclopramide: analysis of hormone profile, progress of oocyte maturation and dependence on temperature. *Aquaculture*. 119, 393-407.

Duncan, D. B., 1955. Mutilple-range and multiple F tests. *Biometrics*. 11, 1–42.

Dye, H.M., Sumpter, J.P., Fagerlund, U.H.M., Donaldson, E.M., 1987. Changes in reproductive parameters during the spawning migration of pink salmon, *Oncorhynchus gorbuscha*. *Journal of Fish Biology*. 29, 167-176.

Epp, R.W., Lewis, Jr. W.M., 1984. Cost and speed of locomotion for rotifers. *Oecologia*. 61, 289-292.

Faramarzi, M., Abbaspour, K. C., Schulin, R., and Yang, H., 2010. Modelling blue and green water resources availability in Iran. *Hydrological Processes*, 23 (3), 486-501.

Faulk, C.K., Benninghoff, A.D., Holt, G.J., 2007. Ontogeny of the gastrointestinal tract and selected digestive enzymes in cobia *Rachycentron canadum* (L.). *Journal of Fish Biology*. 70, 567-583.

Feifei, H., Weiliang, W., Yuanguang, L., Guomin, S., Minxi, W., Jun, W., 2013. Changes of biomass, lipid content and fatty acids composition under a light–dark cyclic culture of *Chlorella pyrenoidosa* in response to different temperature. *Bioresource Technology*. 132, 182–189.

Felix, S. 2013. Advanced growout systems (Raceways and lined ponds) for intensive rearing of ornamental fishes of India. *In*. Ornamental Aquaculture, Technology and Trade in India. Publ. Daya Publishing House, A div. of Astral International pvt. Ltd. 115-126.

Felix, S., 2008. "Biosecured Aquaculture Principle and Prototype" AGROBIOS (INDIA) 1-139.

Felix, S., 2009. Developing Ornamental Aquaculture (Aquariculture) Present Scenario and Scope in Tamil Nadu. *Fishing Chimes*. 29, 8-9.

Felix, S., Herald R.P., Rajeev, A., 2004. Immune enhancement assessment of diatery incorporated marine alga *Sargassum wightii* (*Phaeophyceae Punctariales*) in tiger shrimp *Penaeus monodon* (Crustacia/Penaeidae) through prophinoloxidase (proPO) systems. *Indian Journal of Marine Sciences*. 33 (4), 361-364.

Felix, S., Selvaraj, S., Cheryl A., Moses, T.L.S.S., 2017. First report on induced spawning of Indian pearlspot (*Etroplus suratensis*). *Journal of Entomology and Zoology Studies*. 5 (6), 709-713.

Fernado, A.A., Phang, V.P.E., 1985. Culture of Guppy, *Poecilia reticulata* in Singapora. *Aquaculture*. 51, 49-63.

Fitzpatrick, M.S., Kraak, G.V., Schreck, C.B., 1986. Profile of plasma sex steroids and gonadotropin in Coho salmon, *Oncorhynchus kisutch*, during final maturation. *General and Comparative Endocrinology*. 62, 437-451.

Francis, T., Ramanathan, N., Athithan, S., Cheryl, H.F., 2000. Induced breeding of murrel, *Channa striatus* using various inducing agents. *Fishing Chimes*. 19, 119-121.

Frolov, A.V., Pankov, S.L., Geradze, K.N., Pankova, S.A., 1991. Influence of salinity on the biochemical composition of the rotifer *Brachionus plicatilis* Müller. Aspects of adaptation. *Comparative Biochemistry and Physiology A*. 99, 541-550.

Gao, J., Zhang, H.J., Yu, S.H., Wu, S.G., Yoon, I., Quigley, J., Gao, Y.P., Qi G.H., 2014. Effects of yeast culture in broiler diets on performance and immunomodulatory functions. *Poultry Science*. 87, 1377-1384.

Geist, J., Hawkins, S.J. 2016. Habitat recovery and restoration in aquatic ecosystems: current progress and future challenges. *Aquatic Conservation: Marine and Freshwater Ecosystem*. 26, 942-962.

Gerstner, C.L., Ortega, H., Sanchez, H., Graham D.L., 2006. Effects of the freshwater aquarium trade on wild fish populations in differentially-fished areas of the Peruvian Amazon. *Journal of Fish Biology*. 68 (3), 862-875.

Gilbert, J.J., 2017. Morphological Variation and Its Significance in a Polymorphic Rotifer, Environmental, Endogenous, and Genetic Controls. *BioScience*, 20 (10), 1-13.

Gioacchini, G., Bizzaro, D., Giorgini, E., Ferraris, P., Sabbatini, S., Carnevali, O. 2010. P-230 Oocytes maturation induction by *Lactobacillus rhamnosus* in *Danio rerio*: in vivo and in vitro studies. *Human reproduction* (Oxford, England). 25 (1), 205-206.

Gomez-Gill, B., Roque, A., Turnbull, J.F., 2000. The use and selection of probiotic bacteria for use in the culture larval aquatic organisms. *Aquaculture*. 199, 259-270.

Gonzalez, L.E. and Bashan, Y., 2000. Growth promotion of the microalgae *Chlorella vulgaris* when coimmobilized and cocultured in alginate beads with the plant growth-promoting bacteria *Azospirillum brasilense*. *Applied Environmental Microbiology*. 66, 1537-1541.

Gorelsahin, S., Yanar, M., Kumlu, M., 2018. The effects of stocking density, Tubifex feeding and monosex culture on growth performance of guppy (*Poecilia reticulata*) in a closed indoor recirculation system. *Aquaculture*. 493, 153-157.

Goswami, T.K., Goswami, M.M., 2006. Icthyofaunal diversity and catch statistics ofJamlai wetland in Kamrup District of Assam, India. *Journal of Inland fisheries Society, India*. 38 (2), 38-44.

Goudie, Cheryl, A., Terrence, R., Tiersch, Bill, A., Simco, Kenneth, B., Davis, Qinghua L., 1992. Early growth and morphology among hybrids of *Ictalurid catfishs. Applied Aquaculture.* 3 (3/4), 235-256.

Gouvaia, L., Rema, P., Pereira, O., Empis, J., 2003. Colouring ornamental fish (*Cyprinus carpio* and *Carrassius auratus*) with microalgal biomass. *Aquaculture Nutrition.* 9, 123-129.

Gupta, S., Banerjee, S., 1993. Studies on reproductive biology of *Mystus aor* (Ham.-Buch.,1822), a freshwater catfish of West Bengal, India. *International Journal of Aquatic Biology.* 1 (4), 175-184.

Hamilton, F., 1822. An account of the fishes found in the river Ganges and its branches. *In.* Edinburgh & London. An account of the fishes found in the river Ganges and its branches., i-vii, 1-405, 1-39.

Harvey, B., Carolsfeld, J., 1993. *Induced breeding in tropical fish culture,* International Development Research Centre, Ottawa. 144 pp.

Hess, J.L. and Tolbert, N.E. 1967. Glycolate Pathway in Algae. *Plant Physiology.* 42 (3), 371-379.

Hino, A., 1993. Present culture system of rotifer, *Brachionus plicatilis* and the function of microorganisms. *Finfish Hatchery in Asia.* Proceedings of finfish hatchery in Asia' 91, Tungkang Marine Lab., Taiwan Fisheries Research Institute Taiwan and the Oceanic Institute, Hawai, USA, PP, 51-59.

Hirata, I., 1989. Feed types and method of feeding. *In,* Fukusho, K. and K. Kitajima (eds.), *A Live Feed-The Rotifer, Brachionus plicatilis,* Koseisha-Koseikaku, Tokyo, pp. 73–86.

Hitzfelder, G.M., Holt, G.J., Fox, J.M., McKee, D.A., 2006. The effect of rearing density on growth and survival of cobia, *Rachycentron canadum,* larvae in a closed recirculating aquaculture system. *Journal of World Aquaculture Society.* 37, 204-209.

Holm, J.C., Refstie, T., Bo, S., 1990. The effect of fish density and feeding regimes on individual growth rate and mortality in rainbow trout (*Oncorhynchus mykiss*). *Aquaculture.* 89, 225-232.

Hossain, M.Y., Ahmed, Z.F., Al-Kady, M.A.H., Ibrahim, A.H.M., Ohtomi, J., Fulanda, B., 2008. Threatened Fishes of the World: Wallago attu (Bloch, M. E. and Schneider, J. G., 1801) (Siluriformes: Bagridae). *Environmental Biology of Fishes.* 82, 277-278.

Hotos, G.N. 2002. Selectivity of the rotifer *Brachionus plicatilis* fed mixtures of algal species with various cell volumes and cell densities. *Aquaculture Research.* 33, 949-957.

Houssay, B.A., 1931. Action sexuelle de l'hypophyse sur les poissons et les reptiles. *Comptes Rendus Societe Biologie, Paris.* 106, 377-378.

Hsieh, C.H. and Wu, W.T., 2009. Cultivation of microalgae for oil production with a cultivation strategy of urea limitation. *Bioresourse Technology.* 100 (17), 3921-3926.

Hultberg, M., Jonsson, H.L., Bergstrand, K.J., Carlsson, A.S. 2014. Impact of light quality on biomass production and fatty acid content in the microalga *Chlorella vulgaris*. *Bioresource Technlogies*. 159, 465-467.

Hunter GA, Donaldson EM, Dye HM (1981) Induced ovulation in coho salmon (Oncorhynchus kisutch).

Hunter GA, Donaldson EM, Dye HM (1981) Induced ovulation in coho salmHunter, G.A., Donaldson, E.M., Dye, H.M., 1981. Induced ovulation in coho salmon (*Oncorhynchus kisutch*). Further studies on the use of salmon pituitary preparations. *Aquaculture*. 26, 117–127.

Hunter, G.A., Donaldson, E.M., Dye, H.M., 1981. Induced ovulation in coho salmon (*Oncorhynchus kisutch*). Further studies on the use of salmon pituitary preparations. *Aquaculture*. 26, 117-127.

I. Further studies on the use of salmon pituitary preparations. Aquaculture 26:117–127

I. Further studies on the use of salmon pituitary preparations. Aquaculture 26:117–1Islam, M., Bithy, K., Miah, M., Haque, M., Hasan, K., 2013. Estimation of the Fecundity of Jat Puti, *Puntius sophore* (Hamilton). *Journal of Environmental Science and Natural Resources*, 5 (2), 295-300.

Ibrahem, M.D., 2015. Evolution of probiotics in aquatic world: Potential effects, the current status in Egypt and recent prospectives. *Journal of Advanced Research*. 6 (6), 765-79.

Islam, M., Bithy, K., Miah, M., Haque, M., Hasan, K., 2013. Estimation of the Fecundity of Jat Puti, *Puntius sophore* (Hamilton). *Journal of Environmental Science and Natural Resources*. 5 (2), 295-300.

IUCN, 2018. *The IUCN Red List of Threatened Species. Version 2018-2*. http://www.iucnredlist.org. Downloaded on 14 November 2018.

J. bio-sci. 14: 99-102, 2006 ISSN 1023-8654

J. bio-sci. 14: 99-102, 2006 ISSN 1023-865Jamorz, M., Kucharczyk, D., Hakuc-Btazowska, A., Krejszeff, S., Kujawa, R., Kupren, K., *et al*. 2008. Comparing the effectiveness of Ovopel, Ovaprim, and LH-RH analogue used in the controlled reproduction of IDE, *Leuciscus indus* (L.). *Arch Fish*. 16 (4), 363–370.

ıhedi, A., Jaferian, A., Albooshoke, S.N., 2012. The Effect of Density on Growth and Survival of the Goldfish (*Carassius auratus*, Bloch, 1783). *World Journal of Agricultural Sciences*. 8 (4), 375-377.

ns, L., 1842. The zoology of the voyage of H.M.S. Beagle, under the command of Captain Fitzroy, R.N., during the years 1832 to 1836. Part IV. Fish. - pp. i-xv [= 1-15], [1], 1-172, Pl. 1-29. London. (Smith, Elder & Co.).

Barat, S., 2005. The effect of stocking density on growth, survival rate, and ımber of marketable fish produced of koi carps, *Cyprinus carpio* vr. koi, in ıcrete tanks. *Journal of Applied Aquaculture*. 17, 89-102.

Juchno, D., Boron, A., Gołaszewski, J., 2006. Comparative morphology and histology of the ovaries of the spinedloach *Cobitis taenia* and natural allopolyploids of *Cobitis* (cobitidae). *Journal of Fish Biology*. 70, 1392-1411.

Kagawa, H., Young, G., Nagahama, Y., 1984. *In Vitro* Estradiol 47p and Testosterone Production by Ovarian Follicles of the Goldfish, *Carassius auratus*. *General and comparative endocrinology*. 54, 139-143.

Kaiser, H., 1998. Water quality fluctuations in a closed recirculating system for the intensive culture of the guppy, *Poecilia reticulata* (Peters). *Aquaculture Research*. 29, 611-615.

Kato T., Ohara S., Fukushima Y., Sugimoto A., Masuda T., Yasuhara T., Yamagishi H., 2016. Development of flocculent *Saccharomyces cerevisiae* strain GYK-10 for the selective fermentation of glucose/fructose in sugar mills. *Journal of Bioscience and Bioengineering*. 122 (1), 58-63.

Katsuda T., Shimahara K., Shiraishi H., Yamagami K., Ranjbar R., Katoh S. 2006. Effect of flashing light from blue light emitting diodes on cell growth and astaxanthin production of *Haematococcus pluvialis*. *Journal of Bioscience and Bioengineering*. 102, 442-446.

Kaufman, H., 1965. So you want to raise good guppies! *The Aquarium*, 34 (3), 36-37.

Kelly, J., 1987. *A complete guide to setting up Aquarium* TFH. Publication Inc. New Jersy, p 125.

Kennari, A.A., Ahmadiferd, N., Serfabadi, J., Kapourchali, M.F., 2008. Comparison of growth and fatty acids composition of freshwater rotifer, *Branchionus calyciflorus* pallas, Fed with two types of microalgae at different concentrations. *Journal of World Aquaculture Society*. 39, 235-242.

Khalili, A., Najafpour, G.D., Amini, G., *et al*. 2015. Influence of nutrients and LED light intensities on biomass production of microalgae *Chlorella vulgaris*. *Biotechnology and Bioprocess Engineering*. 20 (2), 284-290.

Khan, H., 1938. Ovulation in fish (Effect of administration of anterior lobe of pituitary gland. *Current Science*. 7, 233-234.

Khoeyi, Z.A., Seyfabadi, J., Ramezanpour, Z., 2012. Effect of light intensity and photoperiod on biomass and fatty acid composition of the microalgae, *Chlorella vulgaris*. *Aquaculture International*. 20 (1), 41-49.

Kim, D-H and Austin, B., 2008. Characterization of probiotic carnobacteria isolated from rainbow trout (*Oncorhynchus mykiss*) intestine. *Letters in applied microbiology*. 47, 141-147.

Kim, J., Sebring, A., Esch, J.J., Kraus, M.E., Vorwerk, K., Magee, J., Carroll, S.B., 1996. Integration of positional signals and regulation of wing formation and identity by *Drosophila vestigial* gene. *Nature*. 382 (6587), 133-138.

Kimura, K., Tomaru Y., 2013. A unique method for culturing diatoms on agar plates. *Plankton Benthos Resources*. 8 (1), 46-48.

Klok Anne, J., Martens, D.E., Wijffels, R.H., Packo, P.L., 2013. Simultaneous growth and neutral lipid accumulation in microalgae. *Bioresource Technologies*. 134, 233-43.

Kobayashi, M., Nakanishi, T., 1999. 11-ketotestosterone induces male-type sexual behavior and gonadotropin secretion in gynogenetic crucian carp, *Carassius auratus langsdorfii*. *General and Comparative Endocrinology*. 115 (2), 178-187.

Kozlowski, B., 1994. Practice of hormonal stimulation of artificial spawning in cyprinid fish. *Broszura Instytutu Rybact wa Srodladowego*. 162, 1-41.

Kreiberg, H., Hunter, G.A., Donaldson, E.M., Craig Clarke, W., Ian Baker, 1987. Induced ovulation and spermiation in the Pacific herring (*Clupea harengus pallasi*) using salmon pituitary preparations and a synthetic gonadotropin-releasing hormone analogue. *Aquaculture*. 61 (2), 155-161.

Kubokawa, K., 1999. Effects of acute stress on plasma cortisol sex steroid hormone and glucose levels in male and female sockeye salmon during the breeding season. *Aquaculture*. 172: 335-349.

Kucharczyk, D., Kujawa, R., Mamcarz, A., Targonska-Dietrich, K., Wyszomirska, E., 2005. Induced spawning in bream (*Abramis brama L.*) using pellets containing GnRH. *Czech Journal of Animal Science*. 50, 89-95.

Kumar, P., Biswas, G., Ghoshal, T.K., Kailasam, M., Vijayan, K.K., 2018. Embryonic and larval developments of brackish water catfish, *Mystus gulio* (Hamilton and Buchanan, 1822) induced with human chorionic gonadotropin and consequent larval rearing. *Aquaculture Research*. 49 (7), 2466-2476.

Kumar, S., Pant, S.C., 1988. Comparative sublethal ovarian pathology of some pesticides [aldrin, methoxychlor and monocrotophos] in the teleost, *Puntius conchonius* and *Puntius gonionotus*. *Bulletin of Environmental Contamination and Toxicology*. 41(2), 227-232.

Kurup, B.M., 1993. Preliminary observation on river ranching of *Labeo dussumieri* (Val.) in Pamba, Kerala. Proceedings of the Fifth Kerala Science Congress, January pp.197-199.

Kurup, B.M., Radhakrishnan, K.V., Manojkumar, T.G. 2004. Biodiversity assessment of freshwater fishes of Kerala with special reference to endemism, threats and conservation measures, *in*, Proc. Second International Symposia on larger rivers. Cambodia. Food and Agricultural Organization (FAO), Rome.

Lavens, P., Sorgeloos, P., 1996. *Manual on the Production and Use of Live Food for Aquaculture*, FAO Fisheries Technical Paper No. 361, 7–76.

Lee, J.S., Newman, M.E., 1997. Aquaculture, 2nd edn. Interstate Publishers, Inc, Illinois II, USA, 393-432.

Legendre, M., 1986. Seasonal changes in sexual maturity and fecundity and HCG-induced breeding of the catfish, *Heterobranchus longifilis* (Clariidae), reared in Ebrie Lagoon (Ivory Coast). *Aquaculture*. 55, 201-213.

Legendre, M., Subagja, J., Slembrouck, J., Hung, L.T., 1999. Larval rearing of an Asian catfish *Pangasius hypophthalmus* (Siluroidei Pangasiidae): analysis of precocious mortality and proposition of appropriate treatments. *Aquatic Living Resources*. 12, 37-44.

Liang, Y., Sarkany, N. and Cui, Y., 2009. Biomass and lipid productivities of *Chlorella vulgairis* under autotrophic, heterotrophic and mixotrophic growth conditions. *Biotechnology Letters*. 31 (7), 1043-1049.

Liang, Y., Xuxin, L., Yuanqi M., Lulu L., Jiaxin,Y., 2018. Interactive effects of microcystin and ammonia on the reproductive performance and phenotypic traits of the rotifer *Brachionus calyciflorus*. *Ecotoxicology and Environmental Safety*. 147, 413-422.

Liley, N.R., Tan, E.S.P., 1985. The induction of spawning behaviour in *Puntius gonionotus* (Bleeker) by treatment with prostaglandin PGF2 alpha. *Journal of Fish Biology*. 26 (5), 491-502.

Lim, L.C., Wong, C.C., 1997. Use of the rotifer, *Brachionus calyciflorus* Pallas, in freshwater ornamental fish larviculture. *Hydrobiologia*. 358 (1/3), 273-275.

Litchman, E., Klausmeier C.A., 2001. Competition of phytoplankton under fluctuating light. *American Naturalist*. 157, 170-87.

Loka, J., Philipose, K.K., Sonali, S.M., 2016. Effect of microalgal diets on filtration and ingestion rates of the rotifer *Brachionus plicatilis*. *Indian Journal of Fisheries*. 63 (3), 75-79.

Lotfi, A., Alain, D., Reynolds, C.S., 2011. Endogenous regulation of the growth-rate responses of a spring-dwelling strain of the freshwater alga, Chlorella minutissima, to light and temperature. *European Journal of Protistology*. 47, 239-244.

Lowry, O.H., Rosebrough, N.J., Farr, A.L., Randall, R.J., 1951. Protein measurement with the folin phenol regent. *The Journal of Biological Chemistry*. 193, 265-275.

Lubzens, E., 1987. Raising rotifers for use in aquaculture. *In* L. May, R Wallace & A. Herzig (eds), Rotifer Symposium IV. *Developments in Hydrobiology* 42. Dr W. Junk Publishers, Dordrecht, 245–255. Reprinted from *Hydrobiologia*. 147-152.

Lubzens, E., Zmora, O., Barr, Y., 2001. Biotechnology and aquaculture of rotifers. *Hydrobiologia*. 446, 337-353.

Ma, E. L., Choi, Y. J., Choi, J., Pothoulakis, C., Rhee, S. H., and Im, E., 2015. The anticancer effect of probiotic *Bacillus* polyfermenticus on human colon cancer cells is mediated through ErbB2 and ErbB3 inhibition. *International Journal of Cancer*. 127, 780-790.

Mahapatra, B.K., 1999. Ornamental fish culture. Freshwater Fishes Research Station. Directorate of Fisheries. Govt. of West Bengal, Kulia, Publication No. 10, p16.

Mahapatra, B.K., 2004. Indigenous ornamental fish germplasm of Northeast. Distribution and conservation status. *In* : Bhattacharjya, B.K. and M. Choudhury (eds). *Ornamental Fish Culture and Trade in Northeastern India. CIFRI (ICAR). Barrakpore, Kolkata*. 39-47.

Mahapatra, B.K., 2010. Breeding and rearing of indigenous ornamental barbs *Puntius shanynius* of NE India. In proceeding of Golden jubilee National seminar on diversification of aquaculture through locally available fish species. Organised by CIFE, Kolkata centre on 27-28 Aug. 2010.

Mahapatra, B.K., 2010. Indigenous ornamental fish diversity of West Bengal Conservation and Management for sustainability. 23rd All India Congress of Zoology & National Conference on Conservation and Management of Faunal Resources for CMFRI, Chennai, 20(12), 7-8.

Mangesh, M.B., Felix, S., Mahadevi, Aruna, S., Cheryl, A., Gopalakannan, A. 2017. Induced breeding of African Jewelfish, *Hemichromis bimaculatus* (Gill, 1862), using gonadotropic signaling molecular analogue WOVA-FH. *Journal of Entomology and Zoological Studies.* 5 (6), 2547-2551.

Mangesh, M.B., Mugale, R.R., Pabitra, B., Honnananda, B.R., Vardia, H.K., 2016. Biotechnological Innovations in Aquaculture. *Ambient Science.* 03 (1), 1-6.

Maragelman, E., Lubzens, E., Minkoff, G., 1985. A Modulator System for Small Scale Mass Production of the Rotifer, *Brachionus plicatilis*, pp. 103–143.

Marimuthu, K., Kumar, D., Haniffa, M.A., 2001. Induced spawning of striped snakehead, *Channa striatus*, using Ovatide. *Journal of Applied Aquaculture.* 19 (4), 95-103.

Marimuthu, K., Muruganandam, M., Haniffa, M.A., 2000. Induced spawning of the Indian catfish *Heteropneustes fossilis* (Singhi) using a synthetic hormone, Ovatide. *Fishing Chimes.* 19 (10-11), 105-106.

Martinez, F., Orus, M.I. and Rosenbrough, N.J., 1991. Interactions between glucose and inorganic carbon metabolism in *Chlorella vulgaris* strain UAM101. *Plant Physiology.* 95, 1150-1155.

Mathias, S., Horst, S., 1990. Response of the photosynthetic apparatus during adaptation of *Chlorella* and *Ankistro desmus* to irradiance changes. *Journal of Plant Physiology.* 136 (6), 675-679.

Mayo, A.W., 1997. Effects of temperature and pH on the kinetic growth of unialga *Chlorella vulgaris* cultures containing bacteria. *Water Environmental Resources.* 69 (1), 64-72.

Menon, A.G.K., 1999. Check List- Fresh Water Fishes of India. *Rec. Zool. Surv. India,* Misc. publ., Occas. Pap. No. 175, 366 pp.

Mercier, L., Audet, C., Joel, N., Parent, B., Parish, C.C., Ross, N.W., 2004. First feeding of winter flounder (*P. americanus*) larvae: use of *B. plicatilis* acclimated at low temperature as live prey. *Aquaculture.* 229 (1-4), 273-278.

Michaloudi, E., Papakostas, S., Stamou, G., Nedĕla, V., Tihlař′ıkova′, E., Zhang, W., *et al.,* 2018. Reverse taxonomy applied to the *Brachionus calyciflorus* cryptic species complex: Morphometric analysis confirms species delimitations revealed by molecular phylogenetic analysis and allows the (re)description of four species. *PLoS ONE.* 13 (9), 1-25.

Mills, G., 1990. An overview of ornamental fishes Fresh and, marine water. *In:* Tomey, W.A. (eds.). *Int. Sem. Ornamental Fishes,* 25 July, 1990.

Mitra, K., 2004. Breeding of ornamental fish with special emphasis on selected indigenous species. *In*: Bhattacharjya, B.K. and Choudhury, M. (eds) *Ornamental Fish Culture and Trade in Northeastern India* CIFRI, Kolkata.

Mitra, S.D., Raizda, S., 1986. Gonadal development of major carps through homeopathy: A novel and successful experiment in Haryana. *Fishing Chimes.* 6, 17.

Mitra. K., Suresh, V.R., Vinci, G.K., Mazumdar, N.N., Biswas, D.K., 2006. Biology and fishery of banded gourami, *Colisa fasciata* (Bloch and Schneider 1801) in a flood plain wetland of Ganga River Basin. *Asian Fisheries Science.* 20, 409-423.

Miwa, Y., Takahide, Y., Yumiko, Y., Hisayuki, A., Tsutomu, M., 2012. Effect of high frequency of intermittent light on the growth and fatty acid profile of *Isochrysis galbana. Aquaculture.* 338- 341.

Mock, C.R., Murphy, M.A., Montanez, F., 1970. Techniques for raising penaeid shrimp from egg to post Iarvae. *Proceeding of World Mariculture Society.* 1, 143-156.

Mohapatra, S., Chakraborty, T., Prusty, A.K., Das, P., Paniprasad, K. and Mohanta, K.N., 2012. Use of different microbial probiotics in the diet of rohu *Labeo rohita* fingerlings: effects on growth, nutrient digestibility and retention, digestive enzyme activities and intestinal microflora. *Aquaculture Nutrition.* 18, 1-11.

Mohsenpour, S.F., Richards, B., Willoughby, N., 2012. Spectral conversion of light for enhanced microalgae growth rates and photosynthetic pigment production. *Bioresource Technology.* 125, 75-81.

Mohsenpour, S.F., Willoughby, N., 2013. Luminescent photobioreactor design for improved algal growth and photosynthetic pigment production through spectral conversion of light. *Biosource Technology.* 142, 147-153.

Morais, M.G. and Costa, J.A.C., 2014. An open pond system for microalgal cultivationin, *In: Biofuels from Algae,* A. Pandey, D.-J. Lee, Y. Chisti, and C. R. Soccol, Eds., I; Elsevier, pp. 1.

Motilan, Y., Bedajit, Y., Vishwanath, W., 2013. Induced breeding of carp minnow, *Puntius chola* (Hamilton, 1822), with synthetic hormone WOVA-FH. *NeBIO.* 4 (3), 21-23.

Motilan, Y., Nishikanta K., 2015. Captive breeding of thick-lipped Gourami, *Trichogaster labiosua* (Day) by gradual increasing aquarium water temperature and their early life stages. *International Journal of Scientific and Research Publications.* 5 (3), 1-4.

Motilan, Y., Yumnam, B., Wahengbam, S., 2014. Captive breeding of endangered Barbs *Pethia manipurensis* (Menon *et al,*) by oral delivery of gonadotropic signaling molecular analogue WOVA-FH. *International Journal of Scientific and Research Publications.* 4 (1), 1-4.

MPEDA. 2016. Ornamental fish assistance scheme. Marine Product Export Development Authority (MPEDA), Kochi, Kerala. 122 pp.

Muthukumar, P., Kandeepan, C., 2015. Isolation, identification and characterization of probiotic organisms from intestine of fresh water fishes. *International Journal of Current Microbiology and Applied Science.* 4 (3), 607-616.

Nagahama, Y., 1983. The Functional morphology of teleost gonads. *In*: Hora, W.S. Randall, D.J. and Danaldson, E.M. (eds). *Fish philosophy*. Vol ix, Part B Academic Press. New York. p. 223–276.

Nagahama, Y., 1987. The Functional morphology of teleost gonads. *In*: Hora, W.S. Randall, D.J. and Danaldson, E.M. (eds). *Fish philosophy*. Vol ix, Part B Academic Press. New York. p. 223–276.

Nandeesha, M.C., Bhadraswamy, G., Patil, J.G., Varghese, T.J., Sarma, K., Keshavanath, P., 1993. Preliminary results on induced spawning of pond-raised mahseer, *Tor kudree*. *Journal of Aquaculture in the Tropics*. 8, 55-60.

Nath, K., 2014. Socio-economic Status of Fish Farmers in Arunachal Pradesh, An Analytical Study. *Asian Journal of Research in Business Economics and Management*. 4 (12), 67-76.

Nath, P., Dey, S.C., 2000. *Fish and Fisheries of North Eastern India (Arunachal Pradesh)*, Narendra Publishing House. Delhi. p. 137–138.

Nelson, J.S., 1994. *Fish of the world* (3rd edition) Joha wiley and sons. Inc. New York. p 371–372.

Nhu, C.V., 2004. A comparison of yield and quality of the rotifer (*Brachionus plicatilis* -L-strain) fed different diets under aquaculture conditions, Vietnam. *Asian Fisheries Science*. 17, 357-363.

Nogrady, T., Wallace, R.L., Snell, T.W., 1993. *Rotifera*. Vol. 1. *Biology, ecology and systematics*. SBP Academic, The Hague, PP.142.

Nwadukwe, F.O., Ayinla, A.O., Abby-Kalio, N.J., 1993. Effects of various doses of acetone dried powdered carp pituitary extract and seasons of hatchery propagation of *Heterobranchus longifilis* (Val. 1840) (Pisces: Clariidae). *Journal of Aquaculture in the Tropics*. 8, 333-340.

Odo, G.E., Agwu, J.E., Iyaji, F.O., Madu, J.C., Ossai, N.I., Allison, L.N., 2015. Mass production of rotifer (*Branchionus calyciflorus*) for aquaculture in south-eastern Nigeria. *International Journal of Fisheries and Aquaculture*. 7 (9), 151-159.

Oeie, G., Olsen, Y., 1993. Influence of rapid changes in salinity and temperature on the mobility of the rotifer *Brachionus plicatilis*. *Hydrobiologia*. 255/256, 81-86.

Ofojekwu, P.C., Arimoro, F.O., (2003/2004). Incidence of feeding, growth, and survival of the toothed carp, *Aphyosemion gairdneri* larvaereared on the freshwater rotifer, *B. calyciflorus*. *Tropical Freshwater Biology*. (12,13), 35-43.

Ogawa, T. and Aiba, S., 1981. Bioenergetic analysis of mixotrophic growth in *Chlorella vulgaris* and *Scenedesmus acutus*. *Biotechnology and Bioengineerig*. 23, 1121-1132.

Olaniyi W.A., Omitogun O.G., 2013. Stages in the early and larval development of the African catfish *Clarias gariepinus* (Teleostei, Clariidae). *Zygote*. 1-17.

Oliver, A. and H. Kaiser, 1997. A comparison of growth, survival rate and number of marketable fish produced of swordtails, *Xiphophorus helleri* Heckel (Family

Poeciliidae), between two types of culture systems. *Aquaculture Research.* 28, 215-221.

Onal, U., Topaloglu, G., Sepil, A., 2015. The Performance of Continuous Rotifer (*Brachionus Plicatilis)* Culture System for Ornamental Fish Production. *Journal of Life Sciences.* 9, 207-213.

Ottera, H., 1993. Feeding, growth and survival of Atlantic cod (*Gadus morhua*) Larval reared in replicate plastic enclosure. *Canadian Journal of Fisheries and Aquatic Sciences.* 50 (5), 913-924.

Ozdemir, N., Ciltas, A., 2010. The effect of caloric restriction on the population density of Freshwater rotifers (*Branchionus calyciflorus*). *Turkish Journal of Zoology.* 34, 377-383.

Pandey, A.K., Mohapatra, C.T., Kanungo, G., Sarkar, M., Sahoo, G.C., Singh, B.N., 2002. Ovatide induced spawning in the Indian major carps, *Labeo rohita* (Hamilton Buchanan). *Aquaculture.* 3(1), 1-4.

Panigrahi, A., 2009. Microbial intervention for better fish health in aquaculture: The Indian scenario. *Fish Physiology and Biochemistry.* 33(4), 429-440.

Papandroulakis, N., Divanach, P., Kentuori, M., 2002. Enhanced biological performance of intensive seabream (*Sparus aurata*) larviculture in the presence of phytoplankton with long photophase. *Aquaculture.* 204, 45-63.

Peter, R.E., Sokolowska, R., Nahorniak, C.S., 1986. Comparison of (D-Arg6, Trap7, Leu8, Pro9 Net) luteinising hormone (LHRA) in combination with pimozide, in stimulating gonadotropin release and ovulation in the gold fish *Carrassius auratus. Canadian Journal of Zoology.* 65, 987-991.

Planas, R., Balleste, B., Alvarez, M.A., Rivera, M., Montoliu, S., Galeras, J.A., Santos, J., Coll, S., Morillas, R.M., Sola, R., 2004. Natural history of decompensated hepatitis C virus-related cirrhosis. A study of 200 patients. *Journal of Hepatology.* 40 (5), 823-830.

Ploeg, A., 2013. The status of the ornamental aquatic industry. *Ornamental Fish International Journal.* 72, 11–13.

Plona, J.E., 1962. Breeding of albino guppy. *The Aquarium.* 31 (3), 127-128.

Polanski, A.S., 1991. Soderzhanic I rezvedenine of kvarium nykh ryb (Maintainence and breeding of aquarium fishes). *Moskva–Russia agropromizdat.* 5(1), 383-389.

Prakash, C., Singh K.M.P., Kiran Dube, A.M., Babitha Rani, Dam Roy, S., Raju, K.D., Pawar N.A., 2014. Studies on captive breeding and larval rearing of *Danio Aequipinnatus* (McClelland, 1839). *Indian Journal of Animal Research.* 48 (4), 379-383.

Pringsheim, E.G., 1946. *Pure Cultures of Algae,* Cambridge University Press, London, PP. 119.

Purkayastha, S., Sarma, S., Sarkar, U.K., Lakra, W.S., Gupta, S., Biswas, S.P., 2012. Captive breeding of endangered *Ompok pabda* with ovatide. *Journal of Applied Aquaculture.* 24, 42-48.

Pyle, E. H., et al., 2008. Dynamics of carbon, biomass, and structure in two Amazonian forests, *Journal of Geophysical Research*. 113, 2-20.

Qie, G., Makridis, P., Reitain, K.I., Olsen, Y., 1997. Protein and carbon utilization of rotifers (*Branchionus plicatilis*) in first feeding of turbot larvae (*Scophthalmus*). *Aquaculture*. 153, 103-122.

Rahdari, A., Ahmad G., Mostafa, G., 2014. Spawning Latency Period in Hormonal Induced Reproduction of Snow trout (*Schizothorax Zarudnyi* (Nikolskii, 1897)). *Iranian Journal of Biotechnology*. 12 (1), 1-5.

Rajendiran, A. and Subramanian, P., 2007. Mass Production of Freshwater Rotifer, *Brachionus calyciflorus*, Under Different Diets and Regimes. *Journal of Applied Aquaculture*. 19 (3), 101-111.

Rath, S.C., Sarkar, S.K., Gupta, S.D., Sarangi, N., 2007. Comparative account of induced breeding of Indian major carps with ovaprim, ovatide, wova- FH and carp pituitary extract. *Indian Journal of Animal Science*. 77(10), 1057-1060.

Reitan, K.I., Rainuzzo, J.R., Oie, G., Olsen, Y., 1993. Nutritional effects of algal addition in first feeding of turbot (*Scoplthalmus maximus*) larvae. *Aquaculture*. 118, 257-275.

Rengpipat, S., Rueangruklikhit, T., piyatiratitivorakul, S., 2008. Evaluation of lactic acid bacteria as probiotic for juvenile seabass (*Lates calcalifer*). *Journal of Aquaculture Research*. 39 (2), 134-143.

Riehl, R., Baensch, H.A., 1996. *Aquarium Atlas*. Vol. 3 : Mergus, Germany. Pp. 1103.

Rimmer, M.A., Reed, A.W., Levith, M.S., Lisle, A.T., 1994. Effect of Nutritional enhancement of live food organisms on growth and survival of Barramundi (*Lates calcarifer*) Larvae. *Aquaculture*. 14, 247 – 260.

Roach, J., 2005. *Conservationists name nine new biodiversity hotspots*. National Geographic news, February 2, 2005.

Rojo-Cebreros, A.H., Ibarra-Castro, L., Guerrero-Carlock, E., Sanchez-Tellez, J.L., Alvarez-Lajonchere, L., 2017. Pilot-scale production of the rotifer *Brachionus* sp. under different culture systems. *Revista de Biología Marina y Oceanografía*. 52 (3), 539-549.

Rosenberg, J.N., Oyler, G.A., Wilkinson L., Michael J.B., 2008. A green light for engineered algae, redirecting metabolism to fuel a biotechnology revolution. *Current Opinion in Biotechnology*. 19, 430-436.

Rosten, T., 2009. Oxygen in water - what is favorable for the fish? (In Norwegian, English summary). *VANN 44 nr 4*, pp 367-374.

Routray, P., Dhananjay, K.V., Sarkar, S.K., Sarangi, N., 2007. Recent advances in carp seed production and milt cryopreservation. *Fish Physiology and Biochemistry*. 33 (4), 413-427.

Ryu, H.J., Oh, K.K., Kim, Y.S., 2009. Optimization of the influential factors for the improvement of CO_2 utilization efficiency and CO_2 mass transfer rate. *Journal of Industrial and Engineering Chemistry*. 15 (4), 471-475.

Saha, S., Behera, S., Mandal, A., Patra, P., Mondal, A., 2017. Early larval development stages of nest building dwarf gourami *Trichogaster lalius* (Hamilton, 1822) in laboratory condition. *International Journal of Fauna and Biological Studies.* 5 (1), 82-86.

Sajeevan, M.K., Somvanshi, V.S., 2013. Collection and rearing of marine ornamental fishes from coastal trawl fishery. *Indian Journal of Geo-Marine Sciences.* 42 (4), 476-480.

Salminen, S., Bouley, C., Boutron-Ruault, M.C., Cummings, J.H., Franck, A., Gibson, G.F.R., Isolauri, E., Moreau, M.C., Roberfroid, M., Rowland, I., 1999. Functional food science and gastrointestinal physiology and function. *British Journal of Nutrition,* 80 (01), S147–S171. Sanders M.E. *Probiotics. Food Technology.* 53(11), 67-77.

Sananurak, C., Lirdwitayaprasit, T., Menasveta, P., 2009. Development of a closed recirculating, continuous culture system for microalga (*Tetraselmis suecica*) and rotifer (*B. plicatilis*) production. *Science Asia.* 35, 118-124.

Sandilayan, S., 2016. Occurrence of ornamental fishes, a looming danger for inland fish diversity of India. *Current science.* 110 (11), 2099-2104.

Sands, D.D., 1986. Keeping Aquarium Fishes, *Corydoras Catfish.,* Dec Bee Book U.K.: pp. 98.

Sarkar, U.K., Deepak, P.K., Kapoor, D., Negi, R.S., Paul, S.K., Singh, S., 2005. Captive breeding of climbing perch *A. testudineus* with Wova-FH for conservation and aquaculture. *Aquaculture research.* 36 (10), 941-945.

Sarkar, U.K., Ponniah, A.G., 2000. Evaluation of North East Indian fishes for their potential as cultivable, sport and ornamental fishes along with their conservation and endemic status. Pp. 11–30. *In : Ponniah, A.G. and U.K. Sarkar (eds.).* Fish biodiversity of NE India. *NBFGR – NADP Pub.* 2, 228.

Sarma, S.S.S., Resendiz, R.A., Nandini, S. 2011. Morphometric and demographic responses of brachionid prey (*B. calyciflorus* Pallas and *P. macracanthus* (Daday)) in the presence of different densities of the predator *A. brightwellii* (Rotifera, Asplanchnidae). *Hydrobiologia.* 662, 179–187.

Sarmah, S.K., 2002. Breeding biology and laboratory propagation of *Danio aequipinnatus* on ornamental fish species of Nagaland. *In : Compendivm on Training on breeding and maintenance of ornamental fishes.* Published by Nagaland University, Lumani, pp. 32–36.

Sarmah, S.K., 2003. Collection, culture and maintainence of *Danio aequipinnatus* as ornamental fish species of Nagaland. *In : Compendium on Training on breeding and maintainence of ornamental fishes. Published by Nagaland University.* Lumani, pp 37–40.

Sarmah, S.K., Dey, S.C., 2004. Captive breeding of endemic ornamental fishes of Northeast – A case study of Silver danio, *Danio devario* (Hamilton) *In:* Ornamental Fish Culture and Trade in NE India (Edited Workshop proceeding) Published by UFRI (ICAR) Brrackpore, Kolkata, pp. 68–78.

Sasmita P., 2016. A review on induced breeding in fishes. *International Journal of Bioassays.* 5 (5), 4579-4588.

Saxby, A., Adams, L., Snellgrove, D., Wilson, R.W., Sloman, K.A., 2010. The effect of group size on the behavior and welfare of four fish species commonly kept in home aquaria. *Applied Animal Behavioral Science.* 125, 195-205.

Scott, A.P., MacKenzie, D.S., Stacey, N.E., 1984. Endocrine changes during natural spawning in the white sucker, *Catostomus commersoni.* II. Steroid hormones. *General and Comparative Endocrinology.* 56 (3), 349-359.

Seema, L., Krishan, R.K., Jyoti, S., 2013. Reproductive Cycle and Oogensis of *Puntius Sophore* from Gho-Manhasan Stream of Jammu region, India. *International Journal of Scientific and Research Publications.* 3 (7), 1-4.

Sen, N., 2000. Occurrence, distribution and status of diversified fish fauna of North East India. *In:* Ponniah, A.G. and U.K. Sarkar (eds). *Fish Biodiversity of North Eastern India. NBFGR – NADP.* Lucknow. 2, 228.

Shehadeh, Z.H., Kuo, C.M., Milisen, K.K., 1973. Validation of an *in-vivo* method for monitoring ovarian development in the grey mullet (*Mugil cephalus L.*). *Journal of Fish Biology.* 5, 489–496.

Shih-Hsin, H., Chen, C.Y., Chang, J.S., 2012. Effect of light intensity and nitrogen starvation on CO_2 fixation and lipid/carbohydrate production of an indigenous microalga *Scenedesmus obliquus* CNW-N. *Bioresource Technologies.* 113, 244-52.

Sichula, J., Makasa, M.L., Nkonde, G.K., Kefi, A.S., Katongo, C., 2011. Removal of ammonia from aquaculture water using maize cob activated carbon. *Malawi Journal of Aquaculture and Fisheries.* 1 (2), 10-15.

Singh, A.K., Ansari, A., Srivastava, S.C., Verma, P., Pathak, A.K., 2015. Impacts of invasive fishes on fishery dynamics of the Yamuna River, India. *Agricultural Science.* 5, 813-821.

Singh, S.P., Singh, P., 2015. Effect of temperature and light on the growth of algae species, A review. *Renewable and Sustainable Energy Reviews.* 50, 431–444.

Sinha, M., 1972. Observation of the biology of *Puntius sarana* 9Ham.) of loni reservoir (MP) : Length – weight relationship, food and condition factor. *Journal of Inland Fisheries Society, India.* 4, 122-131.

So, Y.P., Idler, D.R., Truscot, B., Walsh, J.M., 1985. Progesterones, androgens and their glucuronides in the terminal stages of oocyte maturation in landlocked Atlantic salmon. *Salmo salar* . *Journal of Steroid Biochemistry.* 23, 583-592.

Spolaore, P., Joannis-Cassan, C., Duran, E. and Isambert A., 2006. Commercial applications of microalgae. *Journal of Bioscience and Bioengineering.* 101 (2), 87-96.

Sreevalli S., Ekaterina G., Chandan N., Deepa B., Vijayashree A., Vaishali H., Balasubramanian V., Vasanthi R., Neeraj D., Neela D., 2014. Simple and Rapid Method to Determine Antimycobacterial Potency of Compounds by Using Autoluminescent *Mycobacterium tuberculosis. Antimicrobial Agents and Chemotherapy.* 58 (10) 5801–5808.

Sridhar, S., Vijayakumar, C., Haniffa, M.A. 1998. Induced spawning and establishment of a captive population for an endangered fish, *Ompok bimaculatus* in India. *Current Science.* 75, 1066-1068.

Stacey, N.E., Cook, A.F., Peter, R.E., 1979. Ovulatory surge of gonadotropin in the goldfish, *Carassius aurutus. General and Comparative Endocrinology,* 37, 246-249.

Stahl, M.S., 2009. The role of natural productivity and applied feeds in the growth of *Macrobrachium rosenbergii. Proceeding of World Mariculture Society.* 10, 92-109.

Stevens, C.H., Croft, D.P., Paull, G.C., Tyler, C.R., 2017. Stress and welfare in ornamental fishes, what can be learned from aquaculture? *Journal of Fish Biology.* 91 (2), 409-428.

Strickland, J.D.H., Parsons, T.R., 1972. A practical handbook of seawater analysis, *In.* Fishery Research Board, Canada. 310p.

Suantika, G., Dhert, P., Nurhudah, M., Sorgeloos, P., 2000. High-Density Production of the Rotifer *Brachionusplicatilis* in a Recirculating System, Consideration of Water Quality, Zootechnical and Nutritional Aspects." *Aquacultural Engineering.* 21, 201-214.

Suantika, G., Dhert, P., Sweetman, E., O'Brien, E., Sorgeloos, P., 2003. Technical and Economical Feasibility of a Rotifer Recirculation System. *Aquaculture.* 227, 173-89.

Sudha, C., Gokula, V., 2015. Induced spawning of rosy barb *Puntius conchonius* by *Natrum muriaticum. International Journal of Applied Pure Biology.* 4 (4), 801-804.

Sudhakar, K., Premalatha, M., 2012. Micro-algal technology for sustainable energy production, state of the art. *Journal of Sustainable Energy and Environment.* 59-62.

Sukumaravin, N., Leelapatra, W., Lean, E. Mc., Donaldson, E.M., 2006. Orally induced spawning of Thai carp *Puntius goninotus* (Bleeker) following co-administration of des Gly10 (D-Arg6) s GnRH Ethylamide anddomperidone. *Journal of Fish Biology.* 40 (3), 477-479.

Susan, S.K., Daniel, A.K., Scott, G.L., Clyde, E.G., Lazaro, H., 1998. COMBO, a defined freshwater culture medium for algae and zooplankton. *Hydrobiologia.* 377, 147-159.

Swain, S.K., 2008. Feeding for breeding and maintenance of ornamental fish. *In*: Ornamental fish breeding and culture for income generation. CIFA, Bhubaneswar, Feb.13-19: 56-61.

Swain, S.K., Chakrabarty, P.P., 2008. Indigenous ornamental fish and their export potential. *In*: Ornamental fish breeding and culture for income generation. CIFA, Bhubaneswar, Feb. 13-19, 41-47.

Swain, S.K., Jena, J.K., Ayyappan, S., 2003. Prospects of export–oriented freshwater ornamental fish culture in India. *Fishing Chimes.* 22 (10-11), 111-114.

Swain, S.K., Singh, S.K., 2008. Breeding of gold fish: An ideal strategy. *In*: Ornamental fish breeding and culture for income generation. CIFA, Bhubaneswar, Feb.13-19: 37-40.

Takashima, F., Weil, C., Billard, R., Crim, L.W., Fostier, A., 1984. Stimulation of spermiation by LHRH analogue in carp. *Bulletin of the Japanese Society for the Science of Fish.* 50, 1323-1329.

Talwar. P.K., Jhingran, A.G., 1991. *Inland fishes of India and adjacent countries.* Oxford & IBH Publishing Co. Pvt. Ltd. New Delhi, pp. 1,366.

Taoka, Y., Maeda, H., Jo, J.Y., Jeon, M.J., Bai, S.C., Lee, W.J., Yuge, K. and Koshio, S., 2006. Growth, stress tolerance and non-specific immune response of Japanese flounder *Paralichthys olivaceus* to probiotics in a closed recirculating system. *Fisheries Science.* 72, 310-321.

Thakur, N.K., Reddy, A.K., 1997. Repeat field trials with new hormonal preparation-Ovatide for fish breeding. Final Report, CIFE, Mumbai, India. 124-129.

Thalathiah, S., 1988. Induced spawning of *Puntius gonionotus*. *In:* Maclean, J.L., L.B. Dizon, L.V. Hosillos (eds.), The First Asian Fisheries Forum, Manila, Philippines, 26-31 May, 1986. *Asian Fisheries Society.* 683-688.

Tibile, R.M., Sawant, P.B., Chadha, N.K., Lakra, W.S., Prakash, C., Swain, S., Bhagawati, K., 2016. Effect of Stocking Density on Growth, Size Variation, Condition Index and Survival of Discus, *Symphysodon aequifasciatus* Pellegrin, 1904. *Turkish Journal of Fisheries and Aquatic Sciences.* 16, 455-462.

Tlusty, M., 2002. The benefits and risks of aquacultural production for the aquarium trade. *Aquaculture.* 205, 203-219.

Treece, G.D., Davis, D.A., 2000. Culture of small zooplanktons for the feeding of fish larva. *Aquaculture.* 701, 1-7.

Udit U.K., Reddy, A.K., Kumar, P., Rather, M.A., Das, R., Singh, D.K., 2014. Induced breeding, embryonic and larval development of critically endangered fish *Puntius sarana* (hamilton, 1822) under captive condition. *Journal of Animal and Plant Sciences.* 24 (1), 159-166.

Umesh, C.G., Sudip, K.B., Dilip, B., Konthoujam, S., Bishnupriya, S., Kimneilam, C., 2012. Fish diversity of North East India, inclusive of the Himalayan and Indo Burma biodiversity hotspots zones, A checklist on their taxonomic status, economic importance, geographical distribution, present status and prevailing threats. *International Journal of Biodiversity and Conservation.* 4 (15), 592-613.

Van D.L., Richard, 2017. *Freshwater fish list* (PDF) (23rd ed.). p. 997. ISSN 2468 -9157.

Villegas , C. T., Millamena, O., Escritor, F. 1990. Food value of Brachionus plicatilis fed three selected algal species as live food for milkfish, *Chanos chanos* Forsskal, fry production. *Aquaculture Research.* 21 (2), 213-220.

Vincent, M., John Thomas, K., 2008. Nuptial colouration and courtship behaviour during induced breeding of the swamp barb *Puntius chola*, a freshwater fish. *Current Science.* 94 (7), 922-925.

Vinci, G.K., 1998. Introduction to aquarium fishes. *Fishing chimes.* 17 (3), 25.

Vishakan, R., 2002. Induced breeding in chosen ornamental fishes using Homeopathy preparation, Ph.D thesis, pp. 29-36.

Vishwanath, W., 2010. *Danio dangila.* The IUCN Red List of Threatened Species 2010, e.T166536A6231727. http,//dx.doi.org/10.2305/IUCN.UK.2010-4.RLTS. T166536A6231727.en.

Vishwanath, W., W., Ng, H.H., Britz, R., Singh, L.K., Shivaji, C and Conway, K. W., 2010. The status and distribution of freshwater fishes of the Eastern Himalaya region ICUN Red List, *In:* Status and distribution of freshwater biodiversity in the Eastern Himalaya, pp. 22–29.

Volschenk, C.M., Tweddle, G.P., Janse van Vuren, J. H., Greenfield, R., 2018. Breeding and embryonic development of *Synodontis zambezensis* (Peters, 1852). *Aquaculture Research.* 49 (9), 3083-3087.

Von Ihering, R., 1937. A Method for Inducing Fish to Spawn. *The Progressive Fish Culturist.* 4: 34, 15-16.

Watanabe, T., Kitajima, C., Fujita, S., 1983. Nutritional values of live food organisms used in Japan for mass propagation of fish, a review. *Aquaculture.* 34, 115-143.

Wen, Z.Y. and Chen, F., 2000. Heterotrophic production of eicosapentaenoic acid by the diatom Nitzschia laevis: effects of silicate and glucose. *Journal of Indian Microbiology and Biotechnology.* 25, 218-224.

Whitman, W.B., De Vos, P., Garrity, G.M., Jones, D., Noel, R., Krieg, N.R., Ludwig, W., Rainey, F.A., Schleifer, K.H., 2009. *Bergey's manual of systematic bacteriology.* 2nd edn, Vol. 3, pp. 25-125.

Wu, H., 2016. Effect of Different Light Qualities on Growth, Pigment Content, Chlorophyll Fluorescence, and Antioxidant Enzyme Activity in the Red Alga *Pyropia haitanensis* (Bangiales, Rhodophyta). *BioMed Research International.* 1, 1-8.

Xu, H., Miao, X. and Wu, Q., 2006. High quality biodiesel production from a microalga *Chlorella protothecoides* by heterotrophic growth in fermenters. *Journal of Biotechnology.* 126, 499-507.

Yamamoto, M., Kurihara, I., Kawano, S., 2005. Late type of daughter cell wall synthesis in one of the *Chlorellaceae, Parachlorella kessleri* (*Chlorophyta, Trebouxiophyceae*). *Planta.* 221 (6), 766-775.

Yamasaki, S., Hirata, H., 1985. Effects of salinity on food conversion rate of rotifer *Brachionus plicatilis. Memoirs of Faculty of Fisheries Kagoshima University.* 34, 37-43

Yan Cheng, Zhang Li, Luo Xingzhang, Zheng Zheng, 2013. Effects of various LED light wavelengths and intensities on the performance of purifying synthetic domestic sewage by microalgae at different influent C/N ratios. *Ecological Engineering.* 51, 24-32.

Yasuda, K. and Taga, N.A., 1980. Mass method for *Artemia salina* using bacteria as food. *La Mer.* 18, 53-62.

Yoshimura, K., Tanaka, K., Yoshimatsu, T., 2003. A Novel Culture System for the Ultra-High Density Production of the Rotifer, *Brachionus rotundiformis*- a Preliminary Report. *Aquaculture.* 227, 165-172.

Yuan, D., Zhan, X., Wang, M., Wang, X., Feng, W., Gong, Y., Hu, Q., 2018. Biodiversity and distribution of microzooplankton in *Spirulina* (Arthrospira) *platensis* mass cultures throughout China. *Algal Research.* 30, 38-49.

Yun, Y.S., Park, J.M., 2001. Attenuation of monochromatic and polychromatic lights in *Chlorella vulgaris* suspensions. *Applied Microbiology and Biotechnology.* 55, 765-770.

Zahran, W. E., Elsonbaty, S. M., Moawed, F. S. M. 2017. *Lactobacillus rhamnosus* ATCC 7469 exopolysaccharides synergizes with low level ionizing radiation to modulate signaling molecular targets in colorectal carcinogenesis in rats. *Biomedicine and Pharmacotherapy.* 92, 384-393.

Zehnder, A., Gorham, P.R., 1960. Factors influencing the growth of *Microcystis aeruginosa* Kütz. Emend. Elenkin. *Canadian Journal of Microbiology.* 6 (6), 645-660.

Zeng, T., Huang, H.Y., Kobayashi, N., Li, J., 2017. Performance of an Activated Carbon-Ammonia Adsorption Refrigeration System. *Natural Resources.* 8, 611-631.

Zohar, Y., Mylonas, C.H., 2001. Endocrine manipulations of spawning in cultured fish, from hormones to genes. *Aquaculture.* 197 (1), 99-136.

Zou, E.M., 2003. Current status of environmental endocrine disruption in selected aquatic invertebrates. *Acta Zoologica Sinica.* 49, 551-565.

www.ingramcontent.com/pod-product-compliance
Lightning Source LLC
Chambersburg PA
CBHW062057160726
48282CB00003B/28
* 9 7 8 9 3 5 4 6 1 6 3 4 1 *